The Observer's Pocket Series

GEOLOGY

Observer's Books

NATURAL HISTORY
Birds · Birds' Eggs · Wild Animals · Zoo Animals
Farm Animals · Freshwater Fishes · Sea Fishes
Tropical Fishes · Butterflies · Larger Moths
Caterpillars · Insects · Pond Life · Sea and Seashore
Seashells · Pets · Dogs · Horses and Ponies · Cats
Trees · Wild Flowers · Grasses · Mushrooms · Lichens
Cacti · Garden Flowers · Flowering Shrubs · Vegetables
House Plants · Geology · Rocks and Minerals · Fossils
Weather · Astronomy

SPORT
Soccer · Cricket · Golf · Coarse Fishing . Fly Fishing
Show Jumping · Motor Sport

TRANSPORT
Automobiles · Aircraft · Commercial Vehicles
Motorcycles · Steam Locomotives · Ships · Small Craft
Manned Spaceflight · Unmanned Spaceflight

ARCHITECTURE
Architecture · Churches · Cathedrals · Castles

COLLECTING
Awards and Medals · Coins · Firearms · Furniture
Postage Stamps · Glass · Pottery and Porcelain

ARTS AND CRAFTS
Music · Painting · Modern Art · Sewing · Jazz
Big Bands

HISTORY AND GENERAL INTEREST
Ancient Britain · Flags · Heraldry · European Costume

TRAVEL
London · Tourist Atlas GB · Lake District · Cotswolds

The Observer's Book of
GEOLOGY

I . O . E V A N S F.R.G.S.

FOREWORD BY
PROFESSOR H. L. HAWKINS
(D.Sc., F.R.S., F.G.S.)

WITH 57 PHOTOGRAPHS
AND 65 DRAWINGS

FREDERICK WARNE
LONDON

© Revised edition
Frederick Warne & Co. Ltd.
London, England
1971
Second Reprint 1974
Third Reprint 1977
Fourth Reprint 1979

ISBN 0 7232 0092 0

Reproduced and printed by photolithography
and bound in Great Britain at
The Pitman Press, Bath
D5900.579

CONTENTS

LIST OF PLATES

7 Current-bedded Bagshot Sands; Dorset. (A1418)
Basaltic pillow lava flow (lower carboniferous); Weston-super-Mare. (A11792)

8 Erosion of millstone grit: harder base in coarse grit capping softer bed which is being eroded more rapidly to form stacks; Doubler Stones, Rombalds Moor. (L580)
Small thrust faults dipping to SE in silurian greywacke and shale (D1388)

9 View across the Weald. (A11619)
Diorama of a Cornish China Clay quarry. (MN565)

10 Sharply-bedded, steeply dipping turbidic sandstones, some flaggy, together with siltstones with dark stripy cleaved shales. (A11471)

11 Beds of Kentish Rag (worked as limestone). Narrow bands of hard chert alternate with soft brown calcareous sand known as 'hassock'; Quarry, Offham. (A10277)
Cliffs in blue lias strata; Kilve. (A11686).

12 Sharp anticline in the coal measures showing the steeper drop in the right (northern limb). An equally sharp syncline underlies this to the right; Saundersfoot, Harlow. (A10950)
Anticline in carboniferous limestone; Chepstow. (A10044)

13 The Needles from the East. (A1843)

14 Monoclinal fold traversed by small thrust of lower inclination. Carboniferous sandstone and shale measures; Dears Door, Broadhaven. (A909)
A fold in old red sandstone; Cobbler's hole, St Ann's head, Pembrokeshire (A11873)

15 Clogan Slab Quarries; Oernant. (A3125)
Banded and striped hornblende and granulitic gneiss and mica schists (Burra Voe, Shetland). (D227)

16 Folds in Locheil psammite (moinain); near Lough Quoich. (D1589)
Fold in lower red sandstone. (A878)

17 Contracted mica-schists and quartz mica-schists; Runaby Head. (NI172)

18 A Dartmoor Granite Tor; Blackingstone Rock. (A10059)

19 Pseudo-columnar whin overlying sandstone; Castlepoint, Dunstanburgh Castle. (A3076)
Northern parts of East Mill Tor show horizontal joints typical of the granite; Dartmoor. (A10430)

20 Columnar jointing. (NI304)
Curved columns of Tholeiitic Lava. (NI292)

21 Basaltic pillow lava flow; Spring Cove. (A11792)
Channel and potholes formed in calcareous

volcanic rocks of the Tintagel Group by a freshwater stream; Foreshore at Trebarwith Strand. (A10326)

22 Dyke; Basalt cutting through schists; Ardna Cross Bay, near Campbeltown.
Waterfall in millstone grit Clyn-Gwyn falls, near Ystradfellte.

23 Mica-schists with quartz Augen; south-west of Craig-Fawr.

24 Mountain scree at Wast Water; Lake District. (A11875)

25 View of Stirling showing meanders of the River Forth. (D1010)

26 Head deposits overlying upper chalk; west of Cuckmere Haven. (A9953)

27 The Northern Malverns from Herefordshire Beacons. (A11099)

28 Contorted Eelwell Limestone, Middle Limestone Group (Lower Carboniferous); on foreshore Scremerston. (A3046)
Fold in Wenlock Limestone Shales; Wren's Hill, Dudley. (A1962)

29 Sharp Edge and Scales Tarn with Corrie. (G. V. Berry)

30 Unconformity of Upper Old Red Sandstone

conglomerate and sandstones on vertical silurian greywackes and shales (D1383).
Stromness flags (middle old red sandstone); north from Brough of Biggin. (D.1550)

31 Detail of a fossil tree—Lepidodendron trunk 26 ft+, standing nearly upright; the Cromford Canal Opencast. (L923)

32 Close view of a colony of Litho-strotion pauciradiale, lying close to the base of Malmerby Scar Limestone.
Lower old red sandstone conglomerate. (D1567)

PREFACE

This book is an indirect result of my suggestion to the publishers that they should extend their excellent series of works on Natural History to cover geological subjects. This led them in their turn to suggest that I should add a volume on this intriguing study to their popular *Observer's Books*. I hope that my attempt to comply will induce the observer, the wayfarer with an appreciative eye for the interest and beauty of the countryside, to direct his attention not merely to the living things about him but to the earth which forms their home.

Many of the photographs are Crown Copyright and are reproduced by courtesy of the Controller of H.M. Stationery Office and of the Geological Survey and Museum, for whose assistance I am glad to make suitable acknowledgements. My acknowledgements are also due to :

Messrs. Gregory Bottley & Co. for the drawing of the Geological Hammer, and Mr. R. B. Fuller and Mr. A. F. Stuart for many of the other drawings.

Professor H. L. Hawkins, D.Sc., F.R.S., F.G.S., of Reading University, for his kind assistance and helpful criticism in the production of the work.

I. O. EVANS

FOREWORD

by H. L. Hawkins, D.Sc., F.R.S., F.G.S.
Professor of Geology, Reading University

Geology, in its widest sense, is concerned with everything in this world. Anybody who takes an intelligent interest in anything is therefore a bit of a geologist. The Chemist deals with the substances of which the world is made; the Physicist studies the forces that are at work on those substances, and the Geographer describes the results of those reactions as they are expressed in the form and characters of the earth. The Biologist investigates the nature and behaviour of the relatively small part of the world's substance that is endowed with life; while the Archaeologist, the Historian and the Psychologist concentrate their inquiries upon the peculiarities of one modern species.

Indeed, when the matter is considered in that way, the Geologist seems to hold a sinecure, for almost all of the world's problems seem to have been farmed out among the other arts and sciences. Conversely, if he is expected to know all about the other branches of learning, his outlook is hopeless. But geologists are among the busiest and happiest of mankind. What is it that they find so fascinating and exhilarating?

Like all other students, geologists know that their first business is to observe. Observing involves a great deal more than mere looking, and more even than just noticing; it demands accuracy in both of these coupled with an intelligent attitude towards the objects seen. One notices things that fit in with one's ideas, and one tends not to notice things that seem meaningless.

Of course, nothing can possibly be really meaningless; appreciation (or lack of it) of the significance of anything depends on the experience and psychological make-up of the observer. Ask a dozen readers, chosen at random, for their opinion as to what was the most important news in a single issue of a daily paper, and you will find that the pages and paragraphs that you habitually skip are of the greatest interest to someone else.

So a geologist is first and foremost an observer. He tries to find a meaning in every thing and every event that comes under his notice. When (as happens more often than not) his ignorance prevents him from appreciating the thing he has seen, he either sets to work to improve his knowledge until the appreciation comes, or else passes on his discovery to someone who has the requisite experience for realizing its interest. There is little risk of boredom in a life spent in a wonderland where everything is steeped in mystery that is not always insoluble by honest inquiry.

Let us consider a very simple example of geological observation. Everybody has seen rain falling: most people have, at one time or another, actually noticed it. Let us proceed to observe it; though our observation must be very casual, or this book will be doubled in size and there will be trouble with the publishers.

In the first place, each drop lands with a sputter and splash; if the ground is loose soil or dust, the particles of earth spurt up into the air and fall into a new position. If this happens on a hill-side, nearly all the rain-splash will be directed down-hill, so that the soil 'creeps' (or rather, hops) one grain at a time from the top to the bottom of the slope. The traffic is all 'one-way', and at the bottom of the valley there may be a river (itself a collection of raindrops). That river will carry the particles of soil for, perhaps, hundreds of miles, downhill all the way. Sooner or later, the river will reach a lake or the sea; and there, in the stagnant water, the mud and silt will settle, forming a film of sediment on the floor below the water.

Thus far our observation has been of a geo-graphical nature, and now we are ready to put a geological idea into it. We can be fairly confident that it has been raining, on and off, for a very long time. It would be absurd to imagine that rain could ever have behaved in any way different from that which we can notice today. It is for ever washing the substance of the land into the sea. But there is still quite a lot of land remaining, and signs of the sea getting silted up are few and very local. Evidently our observation has been incom-plete—we have seen only one aspect of the affair. Somehow and somewhere we must look for evi-dence of processes that negative or reverse the obvious degradation of the land.

Armed with this idea we begin to find signifi-cance in all sorts of places. Spectacular events such as earthquakes and volcanic eruptions show that the earth is not content to suffer passively.

Less sensational things, such as spreads of river-gravel far above the reach of floods, or the existence of rings for mooring Viking ships high up on inaccessible cliffs, begin to fit into a pattern which includes the buckling up of the Himalayas. By shifting the load of material from one place to another, the rain interferes with the earth's equilibrium, and readjustments become inevitable.

That is a lot to get from observing the fall of a raindrop, but we can try another idea, and apply it to the same phenomenon. Suppose we dredge up from the sea-floor some of the silt brought down by a river; and suppose that for political or other reasons we are prevented from entering the country through which that river runs. By careful study of the particles of the silt we can form a fairly accurate picture of the nature of that country. Not only will the grains reveal the composition of the rocks over which the rain has splashed and the river has flowed, but chance fragments of water-logged wood or drifted bones will tell us something of the vegetation and animal-life of the area. Like Sherlock Holmes analysing the mud on a criminal's boots, and so detecting his recent travels, we can reconstruct a picture of the unseen territory by examining the dirt that was rinsed from its surface by the rain and deposited on the sea-floor round its coasts.

That word 'reconstruct' holds the pith of the matter. For a geologist is an historian. Believing in the laws of cause and effect, he tries to find the reason for the things he observes. Realizing, like the historian, that history is still in the making, and that Nature, like 'human nature', follows fairly consistent paths, he tries to interpret the fragmentary records of the past in the light of the events of the present. By making fresh observations, and by checking and counter-checking the 'facts' he thinks he knows, he unravels the thread of the story of the evolution of the stage on which he, for the moment, is an actor.

The history he unfolds is one of awe-inspiring grandeur, with none of the petty chicanery and silly presumption that too often sully the human record. And the key to the cypher in which the world's history is written is observation.

INTRODUCTION

'Who bears affection for this or that spadeful of mud in my garden? Who cares a throb of the heart for all the tons of chalk in Kent or all the lumps of limestone in Yorkshire? But men love England, which is made up of such things.'

Himself a geologist, H. G. Wells would certainly not claim that the rocks are lacking in interest, which his historical works so well display. None the less, geology is unlike the other branches of natural history in that its interest is differently centred. In them, it lies in the individual flower or tree, insect or bird or beast or fish, cloud or star; what the inquiring observer wishes to do is to know how to recognize these and to learn something about them. In geology he will, of course, want to recognize the different types of rock and to know something about them, but here the interest of the science only begins.

This diffusion of interest accounts for the arrangement of the present volume, which in some respects differs from the other 'Observer's Books'. It begins by describing the commoner rocks—those which Professor Shand in his *Useful Aspects of Geology* tersely describes as 'Bedded'. It follows this with notes on the structure of these rocks, their foldings and 'faultings', for these influence the scenery as much as the nature of the rock itself.

Next comes a brief description of the different types of fossil which the bedded rocks contain. To make this clear, however, it was found necessary to precede it by still briefer notes on the various systems, corresponding to different periods of the earth's history, in which geologists classify the rocks.

The remainder of the rocks, those which Professor Shand calls 'Massive', which differ in so many respects from the bedded rocks, are then dealt with separately. With few exceptions they contain no fossils, but they form the chief source of the minerals and other ores. Descriptions of the more important or interesting of these accordingly follow.

Finally comes a brief treatment of the different types of scenery in relation to the rocks. To many students this is the most interesting aspect of geology, and it is hoped that the information given will impart this interest to others.

Geological Terms A few terms demand explanation, as their geological usage may conflict with their everyday meaning. To the geologist, for example, the term *rock* means any sizeable bulk of material which forms part of the earth's crust; it does not imply, as in common speech, that the rock is hard. Technically, clay and peat and quicksand are as much rocks as limestone and granite.

The different rocks are made up of *minerals*—substances of a definite chemical composition. Some rocks consist almost entirely of one mineral; chalk, for example, is an impure form of calcite.

Others contain several; thus granite is chiefly formed of three distinct minerals: quartz, feld-spar, and mica. Most bedded rocks also contain *fossils,* the hardened remains or traces of extinct animals or plants.

Of the two different kinds of rocks, the *bedded* consist of parallel layers, *strata* more or less distinct; they were formed by the hardening of sediments accumulated at the bottom, or on the shores, of a sea or other stretch of water or heaped together by the wind. The *massive* rocks show no trace of bedded structure, and were formed by the cooling and hardening of molten material forced up from within the earth.

The layers in the bedded rocks may be horizontal or vertical, or may slope or be contorted into folds. Their slope, measured in degrees from the horizontal and by compass-direction, is called their *dip;* and this in a folded rock-bed will, of course, vary from place to place. A line at right-angles to the dip is horizontal and is called the *strike* of the rock-bed (these terms may be illustrated by lifting the top of this book while keeping the bottom flat on the table: if it then be regarded as a rock-bed, a line straight down the page would represent the dip, the line of print—or the top of the book—would give the strike). Ignoring mere superficial coverings such as the soil, vegetation and the works of man, the area where any rock-bed comes to the surface is its *outcrop;* only in level country will this correspond with the strike.

As scientific terms go, these are refreshingly simple, having been coined not by professors with a taste for the classical tongues but by practical miners and engineers to meet the everyday needs of their work. They contrast with many other geological terms, and especially with the names of the different systems and of the fossils; these, following the usual practice in such matters, are mostly of Greek and less frequently of Latin origin. A few come from Germany, which once led the world in mining technique.

As far as possible such terms have been avoided, though a few have been necessary. Surely, however, nobody need be dismayed at having to master such words as 'anticline' or 'ammonite'. One of the longest, 'carboniferous', is less familiar than 'antirrhinum' or 'carburettor', but its meaning is more obvious—and it is easier to pronounce and to spell.

Practical Work   This book is written in the hope of helping the observer to observe. He cannot be too strongly reminded, however, that geology is not to be learned from books. It demands first-hand observation of the rocks, and an attempt to apply to them the theories and principles which the book sets out.

Little apparatus is required for this work. Apart from a notebook, the one essential implement is a *geological hammer,* weighing about half a kilogramme, with a straight helve and a head squared at one end and tapering at the other to a horizontal chisel-edge. Domestic hammers are useless, but failing a geological hammer the combination of a coke-hammer and a *cold* chisel might serve.

If specimens are to be collected, a good strong haversack will be needed to carry them, and a supply of small boxes or wrapping paper and labels, for each must be separately packed and clearly identified with details of where it was found.

Suitable specimens to collect include fragments of the rocks themselves (from their unaltered interior rather than their weathered surface) and of the minerals and fossils they contain. Except for rock-salt, which needs to be kept in a corked bottle, they can well be stored in wooden boxes. Considered as objects to collect, they have the disadvantage of being heavy and bulky, but on the other hand they do not deteriorate with time; and though most are of purely technical interest, some —well-formed crystals and complete fossils—are of undeniable beauty or interest. As objects of serious study they are rewarding.

The indoor branches of study are beyond the scope of the present book. They include the microscopic investigation of fossils and of rock-sections ground, by a delicate process, thin enough to be transparent. The determination from the geological maps of the structure of the underlying rocks, and the drawing of illustrative sections, forms a fascinating exercise in deduction. This, like the mathematical investigation of the properties of crystals, involves a special application of solid geometry.

Geology may lack the emotional appeal of the other branches of natural history, the charm of animal or plant and the beauty and mystery of the stars. Yet it has the advantage of taking its followers into the finest scenery, and it has more than a technical interest. It aims at deducing from the rocks the nature of bygone landscapes, and from the fossils the character of the plants which clothed them and of the animals which inhabited them, at expounding the history of our earth from its origin as a planet to the appearance of man. It is thus a link between astronomy and human history, and it has done for our ideas of time what astronomy has done for our notions of space. It has inspired trains of philosophical reflection and has profoundly modified our religous ideas. It has attracted to its study men of outstanding intellect, and it has aroused the minds and stimulated the imagination not only of the man in the street but of his child.

BEDDED ROCKS

Chalk

Occurrence   Chalk must certainly be the best-known rock in England. From the great military manoeuvring-ground of Salisbury Plain, it extends northwards into Yorkshire, westwards into Devon, eastwards to the English Channel, and— broken only by a narrow strip of water—southwards to the Isle of Wight. It gives London's countryside some of its finest scenery, the Chiltern Hills and the North Downs. It builds the 'White Walls of Old England', which the traveller returning from the Continent rejoices to see.

Character   Chalk is easy to recognize and impossible to mistake. When freshly broken it is dazzling white, but exposure to the weather robs it of its purity. Being very soft, it powders easily and forms a white 'streak' on anything it touches. It can indeed be used for writing or drawing like blackboard 'chalk', which is not chalk at all, but a form of *gypsum*.

Exposures in chalk-pits, roadside and railway cuttings and cliffs show that the chalk is a bedded rock. It consists of a mass of parallel layers *(strata)* and splits along these layers in flat slabs; otherwise it smashes when broken into shapeless lumps. Near London it *dips* (slopes) so gently that its layers appear horizontal, but in the North Downs it dips slightly towards the north and in the Chilterns towards the south-east, so that it forms a continuous bed under the Lower Thames —wells and other boreholes confirm this. Elsewhere, however, its layers dip more steeply, and in places they are vertical.

Composition Chalk is a form of *calcite,* a mineral which consists of calcium carbonate. Touched with weak acid, it 'fizzes', giving off bubbles of carbon dioxide (this is a usual test for carbonates, an 'acid bottle' containing dilute hydrochloric acid being part of the systematic geologist's equipment). When strongly heated it parts with its carbon dioxide and becomes quicklime (calcium dioxide), so corrosive a substance that in medieval times it was used as a weapon. Water added to quick-lime converts it into slaked lime (calcium hydroxide) meantime generating a heat so fierce as to throw off clouds of steam.

This process gives chalk its chief industrial value. It is 'burnt' (or kilned) in lime kilns to form quick-lime, and this is quenched (or 'slaked') by jets of water. The slaked lime thus formed is used in making mortar and cement and also serves as a fertiliser, restoring to the soil the calcium needed by growing plants. Lime is, moreover, used in the manufacture of steel, as a flux in metallurgy, and in neutralizing acids. Though too soft to be favoured as a building material, the chalk itself has been used for this purpose.

Powdered and examined under the microscope, chalk reveals itself as consisting of white grains, mixed with fragments of marine shells, and even with complete shells of minute creatures. It strikingly resembles the ooze obtained by dredging the bed of the Atlantic. For this reason it was once thought to have accumulated at the bottom of a former ocean; while it was certainly formed under water, however, there are reasons for supposing that this was not very deep.

The chalk, in its hundreds of metres of thickness, varies in texture and composition. It contains belts of 'chalk rock' and other material hard enough to be used for building. Towards its base it is grey rather than white and contains patches of *marl* (limy clay). On the East Coast are layers of Red Chalk—lumps of which, picked up on the beach, serve as a building material; its colour is caused by the presence of iron.

Chalk Scenery Typical chalk scenery consists of Downland, rolling hills, mostly clothed in a short springy turf, admirably suited for sheep-pasturing. Its rounded slopes are broken by many valleys, for the chalk is readily dissolved by the streams which flow over it; most of them, however, are 'dry valleys', the waterways which formed them having dried up or gone underground. Some of these streams reappear temporarily in unusually wet weather—these are called 'bournes' in the South of England and 'gipsies' in Yorkshire. There is indeed a marked absence of water in the chalk hills, so that the farms and villages are mostly situated at their foot; dew-ponds on the crests attract moisture from the air. The lightness of the soil and the graceful outline of the hills give the chalk country a characteristic delicate beauty.

So thin is the soil that it is easily dug away to disclose the gleaming white chalk below. This fact has been known and taken advantage of from

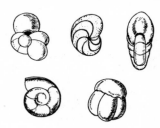

Minute Fossils in chalk (highly magnified)

'time immemorial'. England boasts several ancient figures cut into the chalk: the 'Long Men' (giants) of Wilmington in Sussex, and Cerne Abbas in Dorset, the Whiteleafe Cross near Princes Risborough, Buckinghamshire, the famous 'White Horses'—notably that of Berkshire, really far older than the victory of King Alfred which it is supposed to commemorate—and the chalk-cut figure of George III on White Horse Hill near Weymouth in Dorset. Modern emblems similarly revealing the chalk are the regimental badges cut by the troops stationed on Salisbury Plain during the First World War.

In places, however, the chalk hills are capped with tracts of clay, which encourage the growth of trees. The woodlands of 'Beechy Bucks' grow on the clay of glacial origin (see page 201) which covers so much of the Chilterns. Box Hill, overlooking Dorking, gets its name from its box-trees, and yew and juniper also grow on the chalk.

Flint

Most chalkpits contain *nodules* (rounded lumps) of *flint*. Although most of these form layers parallel to the bedding of the chalk, the two minerals are altogether different in appearance and composition. The flint is very hard and, except on its surface, it is black or very dark grey. It breaks neither into jagged lumps nor along flat surfaces, but in gentle curves like those of an oyster-shell; this is called *conchoidal* (shell-like) *fracture*. Broken into thin flakes, it is translucent and may have very sharp edges.

Composition   Flint is a form of the very common mineral *silica* (silicon dioxide—see page 160). Like the chalk, it is derived from material in the bodies of sea-living creatures. At one time it was supposed that this had accumulated on the seafloor separately from the calcite which forms the chalk; it is now thought more probable that its particles were originally widely scattered through the chalk, but that they gradually 'grew' together by a process resembling crystallization.

The surface of most flints consists of a porous layer of particles of silica so small that by 'scattering' the light they appear white. Broken pieces of flint may 'weather' to a delicate shade of blue.

Flint is used for road-making, in by-roads, as it is too brittle for heavy traffic, and as a building material. In the 'chalk country' many houses and churches are built of flints mortared together. The flints may be split and the broken surfaces turned outwards, forming walls consisting of gentle curves with a slight sheen; or they may be used intact, giving the buildings a 'nubbly' appearance.

Nowadays, flint in small fragments is seldom put to its traditional use of striking sparks for fire-making. Before Humphry Davy invented the Safety Lamp, coal-mines were illuminated by the fitful light of a 'Steel Mill', in which a revolving wheel struck a shower of sparks from a flint edge; these were not hot enough to ignite the dangerous 'fire damp'.

Flint in History Though its modern uses are so limited, flint was for ages the chief raw material —and perhaps the only material—used in industry, and was almost essential to human life. Not only did it serve to make fire, sparks being struck either from two pieces of flint or from flint and ironstone; it was also used to make almost all tools and weapons. Indeed, for the greater part of human history such implements consisted not of steel, not of iron, not of bronze, not even of copper, but of flint.

The wood with which these tools were hafted and on which they were used has long since perished; the more durable flint implements are exhibited in the museums, and may be collected from some river-gravels. They include knives with edges keen enough to serve as razors, saws with well-cut teeth, scrapers for dressing hides, chisels, sharp-pointed borers and awls; they also include spears and axe-heads (arrow-heads were a comparatively recent invention). For these purposes flint was systematically mined; indeed the

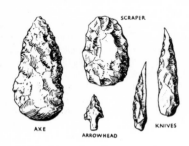

Flint Tools and Weapons

flint-workings of Norfolk have been 'developed' from prehistoric times right down to the present day.

Marcasite Many chalk-pits also contain smaller nodules with a brown, rusty-looking surface: spheres, or cylinders with spherical ends. When broken these reveal a silvery interior with a radial structure, resembling needles tightly packed with their points meeting at the centre; unfortunately, however, the beautiful silver-like surface soon rusts into a reddish-brown. These nodules, which the country folk sometimes call 'Thunderbolts',

consist of iron sulphide, either marcasite or a form
of iron pyrites (see pages 173–4).

Limestone

Character The chalk is found only in the South
and East of England. In many other districts,
however, are rocks which, though very different
from the white chalk in colour and hardness,
nevertheless resemble it in composition. They,
too, consist of calcite, less pure than the chalk and
more discoloured by impurities. They likewise,
'fizz' when touched with weak acid, and they are
burned in kilns to produce lime. For this reason
these rocks are called *limestone*.

The limestones of different regions vary greatly
in colour and hardness. They may be almost
white, creamy, grey, yellow, brown, reddish, or
nearly black; where their calcite is crystalline they
may have a slight lustre. Surfaces exposed to the
weather differ in colour and texture from the rest
of the rock. Some limestones are soft enough to
be sawn (chalk can be cut with a knife); others are
so hard they have to be elaborately hewn with
sledge-hammer and cold chisel.

The limestones also differ in structure. Some,
in the Cotswolds and elsewhere, consist of tiny
grains; their distant resemblance to fish-roes has
earned them the name of *oolite* (from the Greek
for 'roe stones'). Others—as at Leckhampton Hill,
Gloucestershire—consist of larger grains, slightly
flattened from the spherical; these are called

pisolite ('pea-stones'). Many consist largely, or almost completely, of fossils.

The harder limestone, being so resistant to weather as to produce ranges of lofty hills, is called the Mountain Limestone, and because of its proximity to the coal measures, it is also known as *Carboniferous Limestone* (from the Latin for 'coal-bearing'). It helps to form 'the backbone of England' (the Pennine Chain) and builds the Peak and Fells of Derbyshire and the Mendip Hills. It also occurs in Wales, around the Lake District, and in the Scottish Lowlands, and it occupies a wide area in Central Ireland.

Building Stone Like the chalk, the limestone consists of parallel beds. It may be traversed by *joints* (cracks) running at right-angles to the bedding and dividing it into sizeable blocks. *Freestones,* as rocks so divided are called, are deservedly in demand among builders. Many important building-stones are limestone: Kentish Rag, Portland Stone, Caen Stone, Bath Stone, and the so-called Purbeck Marble, which is not a true marble but simply a limestone which can easily be polished. Limestone has thus two great industrial uses: it is a building material as well as a source of lime.

In its pure form the mineral calcite is transparent and colourless, and may form perfectly shaped six-sided crystals (see page 167). It may also be white, or tinted grey, brown, or red by traces of impurities. Its crystals, when wide and squat, are called 'nail-head spar', when longer and narrow, 'dog-tooth spar'.

Dolomite Very similar to calcite is *dolomite* (called after a range of mountains in the Tyrol, itself named after a French geologist), a carbonate of magnesium and calcium; this also forms six-sided crystals, coloured pale brown or yellow and with a pearly lustre. Traces of this mineral occur in many limestones, and it is the chief constituent of dolomitic limestone, also called Magnesian Limestone. This forms a narrow belt running southwards from the Durham coast to the neighbourhood of Nottingham. It produces steep yellow cliffs where it reaches the coast and a steep scarp facing westwards where it runs inland.

Chert To complete the resemblance of limestone to chalk, it may contain nodules of silica. This, however, does not break with so regular a conchoidal fracture, nor is it as translucent as the flint. So different is it from this that it is called *chert*.

Limestone Scenery The limestone, being similar in composition to, though harder than, the chalk, naturally produces a similar but bolder scenery. Some travellers find the limestone country forbidding. Sydney Smith described the Cotswolds as 'one of the most unfortunate desolate counties under heaven' and asserted that after travelling far over it 'life begins to be a burden and you wish to perish'. The only relief he could find was in the river-valleys beyond the Cotswolds, though he might have appreciated the smaller vales which intersect them. Other lovers of scenery, however, find a real delight in the austere wildness which some of the limestone hills display.

The oolite of the Cotswolds and the hills farther north (for it runs north-eastward across England, parallel to the chalk of the Chilterns) forms a broad range of hills, sloping gently upwards from the south-east and terminated along its north-western edge by a steep scarp. The fields are mostly separated not by hedges but by walls built, like many of the houses, of the local stone, harmonizing excellently with the rolling hills and sharing their character. Beyond the scarp are isolated hills consisting of limestone *outliers*, masses of the rock separated from the main outcrop by weather action.

An Outlier

Bleak and forbidding—or attractive, according to the point of view—are the hills formed of the harder Mountain Limestone. Among them are the 'lows' of Derbyshire, so delightfully contrasted with the Dales which intersect them. Like the great Pennine Ridge, they slope more steeply than those formed of the oolite or of the chalk. The scarps which terminate their slopes are more precipitous. Their detached rocks—a striking example is the High Tor of Matlock—are higher and more majestic. Their vales are deeper, and are bordered by abrupt slopes or even by cliffs, their vertical walls gleaming behind the trees.

Limestone Fossils Many limestones are so rich in fossils as to be named after them. The Coral Limestones of the Mendip Hills and the Cotswolds, and the similar 'Corallian Limestones' of the Oxford region are coral reefs deprived of the tiny sea-creatures which built them and risen to form dry land. They also contain fossils of the shellfish which lived among the reefs. The Shelly Limestones of the Cotswolds and Derbyshire teem with the fossils of oysters, lamp-shells, ammonites, and other shellfish. The limestones of Wales, Derbyshire, and North Lancashire are in places so packed with the remains of Sea-lilies (Crinoids) that they are often called 'Crinoidal Marble'. The Algal Limestones contain fossils of marine plants. Other limestones include the fossils of sponges or of microscopic animals or plants. The rock itself may be so worn away by the weather that its fossils stand out in relief.

The limestone may also contain larger fossils, which vary in type according to its 'geological age' (see page 84). Among these are the bones of the giant reptiles which at one time dominated the earth. It may also show indications of other forms of life; hard as it is, certain shellfish and other animals are able to bore holes into the solid rock.

Underground Water This seems unlikely; and no less unlikely does it seem that even the hardest mountain limestone should dissolve in water. Yet 'constant dripping wears away a stone';

37

rain, purest natural form of water though it is, is really a very weak acid, having dissolved carbon dioxide and traces of other chemicals from the air. Weak though the acid is, it is powerful enough to attack the limestone, and imperceptible though this action may seem, its effects are cumulative. It acts irregularly, for the rock varies greatly in hardness and composition. It is aided by the equally irregular action of the rain, the frost and the sunshine. These agencies, continuing over long periods, groove the limestone hills with valleys, detach the outliers from the main escarpment and carve the edges of the cliffs into fantastic shapes.

Moreover, the action of the water takes place underground, for limestone is traversed by joints and the chalk is *permeable ;* loose in texture, it absorbs water like a close-textured flannel or sponge. Seeping down into them, the rain accumulates above whatever layer of *impermeable* (watertight) rocks may be below. Thus prevented from seeping deeper, it saturates the chalk or the limestone. It overflows into surface springs and fills natural rock-crannies and artificial borings and wells. It can, indeed, be compared to an underground lake. Though this rock-held water is a hindrance to the engineer, it is a boon to the neighbourhood, affording a copious water-supply. Professor G. M. Davies aptly says in his book, *Geology of London and South-east England,* that 'Economically, the most important product of the chalk is water'. This might almost equally be said of the limestone.

Water which has dissolved calcite from the rocks is very different from the 'soft' rainwater. It is *hard* water. It lathers only with difficulty, the soap at first combining with the calcite to produce an unpleasant scum. The hardness, unlike that produced by certain other minerals, is only 'temporary' and can be removed by boiling. Dissolved calcite also renders the water very transparent, so that streams flowing from the chalk or limestone are delightfully clear.

Petrifying Springs Hard water leaves a thin stony coating on the inner surface of the utensils in which it is boiled. It leaves a similar coating wherever it evaporates, many springs in limestone country being surrounded by a stony crust known as *travertine* or calcareous (limy) *tufa*. This is taken advantage of by the proprietors of 'petrifying springs' in Matlock and elsewhere, where the water is made to spurt over a miscellaneous collection of objects ranging from tin-hats to birds' nests. For a small payment visitors may see the thin coating of stone-like tufa which gradually covers them; and for a larger fee they may add one of their own possessions to the heap and have it posted on to them later, duly 'petrified', as a curio.

Limestone Caves

Pot-holes and Caves Seeping down through the chalk, or trickling through the crevices of the limestone, the rainwater dissolves the rock. Soon, naturally, it becomes a 'saturated solution', holding as much calcite as it can dissolve, so that its action ceases. Still, however, it is reinforced by further rainwater from above. Long continued, this action pierces the limestone, especially along the joint-places and its bedding, with vertical *swallow-holes* or *pot-holes*, and with horizontal *caverns* or *caves*. Running along the strata of the rock, the caves mostly have level floors, but they vary greatly in width and height—from a mere cranny to a great vault. If it finds a weak place in the rocks beneath, the underground water may cut another vertical chasm through this to resume its horizontal course some distance below.

'Pipes' and Dene-holes The chalk is too soft to allow caves to be thus formed; the layers above at once collapse into them, leaving a depression in the ground. The material above the chalk may soon wash into this depression, forming vertical 'pipes' of sand, gravel or clay; normally invisible, these are revealed in chalk-pits and embankments. The so-called 'caves' in the chalk—the most remarkable of which is Chislehurst Cave—are not natural, but were artificially excavated (not, as romantic guides state, by the 'Druids' as underground 'temples' but simply in much more recent times by prosaic quarrymen engaged on mining chalk). The *dene-holes* of Essex and Kent, vertical shafts sunk in the chalk, with two or three rounded chambers at their foot, were similarly of artificial construction, though their purpose is uncertain.

Cheddar Gorge In the harder limestone, however, the upper layers of the rocks are solid enough to remain intact even when undermined by a complicated series of caves, such as those of the Mendips, of Derbyshire, and of the Ingleborough District. Only on rare occasions does the roof fall in, producing a steep-sided valley or gorge. Cheddar Gorge, that magnificent chasm which cuts through the Mendip Hills from north to south—it is Britain's nearest approach to a canyon—may be the remains of an immense cave whose roof, weakened by the weather, has collapsed; but on the other hand it may represent the former valley of a river whose waters now find a course underground.

The Cheddar Gorge

Underground Streams Some of the streams which formed the swallow-holes are still flowing, pouring over their lip in a waterfall; some of those which formed the caves still vanish into natural tunnels. They may reappear miles away, as is proved by simply casting dye into the streams where they disappear into the earth and noticing where the discoloured water emerges. Many caves, however, are nowadays quite dry, the streams which carved them no longer flowing; and in many others the streams are now represented only by underground pools or lakes in the depressions in their floors.

Cave Exploration Not all of underground Britain has been explored. Such exploration is a branch of science known as *speleology* (from the Greek word for 'cave'). It is an exciting and hazardous sport, which has aptly been described as 'mountaineering in reverse'. The explorers may have to be lowered by ropes *through* a waterfall to the bottom of a swallow-hole; they may have not merely to crawl but even to wriggle through crevices hardly large enough to admit them; they may have to surmount underground precipices; they may have to use folding canoes or portable boats to cross the underground ponds and to traverse the underground rivers; they may have to wade, or even to swim.

For systematic cave exploration an assortment of gear may be needed; not only commonplace ropes and rope-ladders but field-telephones, cork-jackets (a frogman's outfit) or collapsible boats. Serious speleologists take scientific instruments, from geologists' indispensable hammers and collecting-bags to barometers, theodolites and rockets or fire-balloons (for measuring the heights of lofty caves.) It is obvious that amateurs and experts alike need generous food supplies and a plentiful store of lighting materials, supplementing acetylene flares and powerful electric torches by candles and matches carried in waterproof containers. Still more obvious is it that cave exploration is not to be attempted except under experienced leadership.

The more accessible caves are exploited commercially, being lighted electrically; a few are even said to be embellished by stone icicles and the like pillaged from elsewhere and fixed artificially to their roofs! Guides escort parties in return for a fee and a tip, their talk often being more voluble than accurate. Even caves such as these can be dangerous, however, if a visitor unwisely strays from the beaten track.

Prehistoric Cave-drawings These commercialized caves are most impressive, and those still unexploited are awe-inspiring indeed. The darkness around, the wavering shadows produced by the moving lights, the echoes of voice or footfall, the drip of falling water, the general strangeness of the surroundings, give an impression of eerie mystery. It is no wonder that the caves of southern France and northern Spain (though not, regrettably, of Britain) are decorated with drawings and paintings of animals made by primitive man thousands of years ago, with the aim, it is thought, of exercising a magic power over the creatures they represent. (See pages 139–40.)

Dripstone Strangely enough, the waters which

formed the caves begin a process which would end by filling them. Saturated as it is, each water-drop deposits in evaporating a speck of calcite, and these specks, minute as they are, accumulate to build large deposits of that mineral in a form called *dripstone*. The water which trickles down the walls gives them a coating of this waxy-looking material; where it flows over irregularities in the wall it forms sheets which may resemble a frozen cascade or a mesh of delicate filigree-work; where it accumulates on the cave floor it may form a series of equally delicate dykes. Like the other forms of calcite, it may be almost transparent, milky-white, or tinted pink, brown or red with iron.

Stalactites The water which drips from the cave-roofs similarly deposits minute specks of calcite, and these accumulate to form what look like stone icicles hanging from the roof. (Similar stony 'icicles', formed from the lime in the mortar, grow under some railway bridges.) These *stalactites* vary greatly in size and in proportion, but are all graceful and pleasing. (A few of them have their points upturned, a fact very perplexing to the geologist.)

Stalagmites The falling water also deposits its dripstone specks on the cave-floor, and these similarly accumulate. They mostly form not graceful, slender 'icicles' but squat hummocks like waxy stone molehills called *stalagmites* (both terms come from the Greek word for 'drip'). As the drops continue to fall and the calcite specks to accumulate above and below, the 'icicle' lengthens and the hummock grows upwards; at last the two may meet to form a dripstone column running from cave-floor to roof, giving the effect of a pillar in some natural cathedral. Such columns are of varied, even of fantastic, shapes; they add greatly to the attractiveness of a cave, and the guides delight in giving them fanciful names.

Sand and Sandstone

The 'Red Downs' How greatly the scenery depends upon the underlying rocks is illustrated not far south of London. Here two lines of hills run parallel for miles. They are separated only by a narrow valley. They differ little in height and hardly at all in latitude. They are exposed to the same weather conditions and to similar human activities. Yet they differ strikingly in soil and in plant and animal life.

The line nearer to London, the North Downs, has already been described. It consists of rolling slopes of grassland with clumps of trees here and there, among them the yew, juniper and box, with the beech where there is a surface layer of flinty clay. The more southerly line (occasionally called the 'Red Downs' to distinguish it from the White) is a 'forest' in the historical sense of the term, a tract of wild country—is not this part of the 'Weald'?—of pinewoods interspersed with heathland.

A clue to the reason for the difference is given

in the soil, the firm turf of the White Downs contrasting strongly with the brown, sandy earth of the 'Red'. This contrast is itself due to the difference in the rocks from which the soil is formed. A pit in the North Downs reveals the gleaming chalk; a pit in the 'Red Downs' reveals immense layers of sand.

Sand: Character Sand is familiar enough to all who have visited the seaside. It consists of countless loose grains, varying in colour from white to brown or dark grey, and in size from a fine dust to tiny pebbles. The microscope reveals these grains as minute fragments of stone of varying degrees of smoothness or roughness. Chemical analysis shows that most of them consist, like the flint, of silicon dioxide (silica).

The sand is derived from the powdered remains of quartz—another form of silica. First broken up by the weather, the rock fragments are rolled and ground smaller and smaller by the sea, and at last washed up in fine grains upon the shore. Here, when damp, the sand levels out to form the beaches, wide stretches of powdery material hard enough to walk on yet soft enough to take an impression of whatever touches it: imprints of human feet, tracks of walking animal or hopping bird, marks left by crawling sea-creatures or brushing fronds of seaweeds, rippling water or falling rain.

Sand Dunes When dry, the finer sand is readily blown along. The wind piles it up to form the *dunes,* which resemble great sluggish-moving waves: convex on their windward side, concave to their lee and with the concavity topped by steep, almost vertical, crests. Like sluggishly-moving waves they slowly travel across the land.

Wind-blown sand-dunes can be appallingly destructive. They can choke and divert rivers and bury not merely fields and meadows but farms and even whole villages and churches. The former church-tower of Eccles, north of Yarmouth, was at one time partly buried by moving sand. Its remains were exposed when the gales of 1947 blew the sand clear. One of the 'sights' of North Cornwall is the Church of St. Enodoc's. It is surrounded by sand reaching as high as its window-sills and only separated from its walls by an artificial trench dug to the level of its floor. Formerly, it is said, the whole church was almost buried beneath the dunes, so that the vicar, pluckily carrying out his annual service, had to enter and leave, by means of ropes, through a hole in the roof!

Fortunately, sterile as it seems, the sand of the dunes affords enough nourishment to grow such coarse plants as marram grass. Judiciously sown, this grass 'binds' the sand with its roots and prevents it from being blown away. When the grass dies, its remains act as a natural manure, giving nourishment for other hardy plants which continue the binding process. The wide stretches of sand at Braunton, North Devon, show varying stages of dune-reclamation, and foster unusual types of vegetation.

Sandstone The sands revealed in the 'Red

Downs' by quarrying and in the banks of the sunken roads are obviously similar to those of the shore. Though firm enough to form a vertical or steeply-sloping surface, they can be crumbled with the fingers into loose grains. Elsewhere there are other rocks similarly composed of sand-grains, but welded by a natural cement into solid masses hard enough to serve as building material; such rocks are called *sandstone*.

Sandstones vary in colour from white to the yellow, brown or red which indicates the presence of iron, or to the greenish hue which comes from the mineral *glauconite*. In Breconshire and Herefordshire the soil is noteworthy for its dark red hue. This is caused by the presence of iron in the underlying rocks, known as the *Old Red Sandstone*. The brighter red of the soil in parts of Cheshire and Devon is also caused by iron in what is less frequently called the 'New Red Sandstone' below.

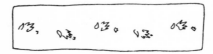

Fossil footprints

The natural cement which binds the sand-grains into solid rock may consist of clay, ironstone, limestone—or itself of silica. Sandstones cemented together by silica are so hard as to demand a special name: *quartzite*. Among the grains of sand may appear specks of the gleaming mineral mica; these are usually in parallel layers, as are the sand-grains themselves. For the sandstone, like the limestone and chalk, is a bedded rock, formed by a sediment which drifted down through the sea or was piled up, layer upon layer, on the beach.

Ripple-marks and Rain-prints The rock splits along its layers, the surface of which may either be level or form undulations, like those produced by the waves on the beach; these are called *ripple-marks*. Other sandstones are marked with *rain-prints,* the tiny pits formed on a bygone sea-beach by falling drops of rain. Others again bear the tracks of crawling sea-creatures, or the foot-prints of animals or birds, or even the marks produced by the tips of swaying seaweed fronds.

Current-bedding In some sandstones the strata run parallel, like those of the limestone or chalk. In others they form a number of parallel curves, varying greatly in direction from place to place. Such an arrangement of varied curves is called *current-bedding*, also false-bedding because it is attributed to currents in the water at the bottom of which the beds were formed; it may be compared to the series of oblique layers formed in a slag-heap as material is tipped over the sloping end of a horizontal mound.

Sandstones, like limestones, may be divided into great rectangular blocks by joints running perpendicular to their layers. They, too, are called *freestones,* and are greatly in demand for building. Both the hard and the loose sandstones are indeed valuable economically, the one as a building material, the other as a source of sand.

Grits The grains in some sandstones are fairly smooth; in others they are so very angular that the rock which contains them demands a special name: it is a *grit*—a term also applied to sandstones with unusually large grains. The *Millstone Grit*, so called because its texture is admirably suited to the grinding of corn, is, like the Mountain Limestone, associated with the coal measures. It produces the wild moorland country to be found along the Pennine Chain and in South Wales. Because of its hardness, it forms bold cliffs—called 'Edges' in the Midlands—and because of differences in hardness between its layers it is readily cut into fantastic shapes by the winds, which drive loose grains against it like a natural sand-blast.

Minerals The Lower Greensands of the Weald are so soft and so readily crumbled that it seems strange they should form some of the highest hills of the South of England and the London region. Soft as they themselves are, however, they are reinforced by beds of chert and of another hard material that is very resistant to the weather. This rock may be seen projecting from the sides of the sunken roads which lead southwards from the Vale of Holmesdale. It rather suggests slabs of irregular rusty boiler plate and is *ironstone,* a form of iron ore.

In the sands near Reculvers, east of Herne Bay, on the north coast of Kent, are crystals of a greenish translucent mineral, called *selenite* (sulphate of calcium—see page 170). This substance has little economic value, and is not specially ornamental; it is of interest only to the geologist or as a curio.

Sandstone Scenery As in the Weald and in the Millstone Grit country, sandstone is not suited to cultivation but produces a wild scenery: common-land, pine-wood and moorland. This is made even more attractive by the dryness of its soil—the rain quickly seeps in between the loosely packed grains, to accumulate below the ground; here it overflows in springs or may be reached by pumps and wells. The rock is not soluble as is the chalk, however, and for this reason is not traversed to any large extent by caves (seashore caves, formed by the beating of the waves, are quite small and do not penetrate far into the sandstone). The dryness of the soil renders the sandstone country greatly in demand as a building site.

Sandstone Fossils The Lower Greensand includes very few fossils. These are, however, to be found in greater abundance in some of the other sandstones found elsewhere in our islands; they are too varied to describe. The Old Red Sandstone of Scotland is noted for its fossil fish.

Silt and Siltstone The very fine grains brought down by the rivers, too small to be regarded as sand, accumulate to form *silt*. Just as sand becomes sandstone, so silt hardens into a similar— though of course more finely-grained—rock, *siltstone*. The grains in this may form thin layers, along which it easily splits.

Conglomerate

Pebble-beaches Just as large grains of sand grade into small pebbles, so these grade into larger pebbles and thence into rounded boulders. These pebbles are less easily moved than the sand; it takes a more powerful current to transport them, piling them up into *shingle-beaches*, which only a storm will demolish. They may form ridges higher than the land immediately behind, and firm enough to dam the streams.

Conglomerate: Description Like the sand and the silt, the pebbles may be welded by a natural cement into solid rock. This is *conglomerate*, also called 'Puddingstone' on account of its appearance. A similar rock formed not of rounded pebbles but of angular rock-fragments, the scree formed by the weather on mountain slopes, is a *breccia* (from an Italian word meaning 'fragments of masonry').

The pebbles in a conglomerate may range from stones so tiny that they resemble a coarse sand to rounded boulders a metre or more across. They may be composed of almost any hard rock—sandstone or flint, for example. They may be arranged higgledy-piggledy or lie regularly in parallel strata; the pebbles in one particular layer may be roughly all the same size; and the pebble layers may alternate with layers of sand. As in the sandstones, the cementing material may be sandy or clayey or consist of limestone or ironstone.

Varieties of Conglomerate The cement may

hold the pebbles so loosely that they easily fall out or can be picked out with the fingers. On the other hand, it may become not merely as hard as the pebbles, but even harder. A loose conglomerate breaks *round* its pebbles, so as to leave them entire; but one formed of hard material may have become so completely unified that it splits more easily *through* the pebbles than round them.

Thus the pebbly rocks at the crest of the North Downs near Caterham are so loosely packed that they could be regarded as an inland shingle-bed rather than as a conglomerate. On the other side of the Thames, however, the pebbles have been welded into a very hard rock, the Hertfordshire Puddingstone. Although the cementing material differs so much in hardness, the pebbles in both consist of the same material, flint. In the Wye

Puddingstone

Valley are slabs of conglomerate so massive that prehistoric man was able to erect them as 'standing stones'.

Clay

Character   A rock with which many of us are unpleasantly familiar is *clay* (in geology, it will be remembered, any material which helps to form the earth's crust, irrespective of its hardness or texture or of common usage, is a 'rock'). The most striking feature is its impermeability to moisture; it is a 'watertight' rock. Rain which falls on a clay

53

soil hardly soaks in at all, but remains on or near the surface, combining with its upper layers to form a slippery mire. In dry weather, the sun-baked clay can become surprisingly hard, and in such countries as Egypt, where rain is unknown, it is used to form bricks.

China Clay   The different clays vary in chemical composition: pure clay is white—the *kaolin* or china clay found in Cornwall. Most clays are, however, coloured grey, brown, reddish, or less frequently dark blue, purple, or green, by impurities; they may include sandy grains of quartz, which gives them a gritty feel. They consist in varying proportions of oxides of iron and aluminium, and of compounds of these metals with silica.

Uses of Clay   Though unsuited in our rainy and temperate climate for building even when sun-dried, clay can be readily 'fired' to make our traditional building material, brick. Certain clays have their special uses; the pure kaolin of Cornwall, as its alternative name implies, is the raw material of china and porcelain. It is also used in pencil and paper making and—of all unlikely uses —as a component of chocolate-cream; the quarry-men even drink clay-water to prevent indigestion! Clays not quite so pure as the kaolin are known from their uses as *potter's clay* and *pipe-clay*. The *fire-clay* found in the coal measures similarly gets its name because, being very resistant to heat when burned, it forms the fire-bricks needed, for example, in furnaces.

Among its other uses, clay is essential in

metallurgy. It forms a component of cement and acts as a 'binder' of road surfaces. It makes not only bricks and tiles but sanitary ware and filters. It helps to size cotton and cleanse wool and to give metal polishes and cleaning powders their abrasive effect.

Clay in History Though admittedly neither so old nor so essential to human life as that of flint, the use of clay is very ancient and very important. What is more, clay, unlike flint, still serves its traditional purposes. Pottery is older than civilization; and the most antique potsherd, like most modern teacups, is made of baked clay. Naturally dried in the sunshine, or artificially baked, clay makes, as it has done through the years, the bricks and tiles for use in building and roofing houses. In the absence of soap, clay can be used, as it was before soap was invented, for personal cleansing. In times of extreme famine, certain types of clay have been used, even within living memory, as a substitute for food.

Perhaps the most interesting use of clay was as a writing material. The *cuneiform* ('arrowlike') writing of Mesopotamia was made by pressing a bone 'style' with a wedge-shaped end against a tablet of clay; this was then dried in the sun or baked, and might be inserted in a clay 'envelope' similarly inscribed. It seems appropriate that clay should still serve in paper-making to fill the minute holes between the fibres of pulped cloth, in pencil-making to hold together the particles of black-lead, and in pigment-making to give body to certain colours.

London Clay The brown clay which forms the floor of the lower Thames Valley is known in Britain as London Clay (on the farther shore of the English Channel it continues inland to the Seine, and is known in France as Paris Clay). Its brown colour, however, is only superficial, and is due to the iron it contains having 'rusted' in contact with the air; when dug up it is bluish-grey. The clay contains traces of several minerals. Near Reculver, east of Herne Bay, may be found large crystals of selenite. It also contains concretions (hard, rounded lumps) of claystone over thirty centimetres across.

Clay Scenery The London Clay forms a wide stretch of low-lying country, almost flat or gently undulating. It constitutes a good pasture-land, and is famed among gardeners for growing roses; elms and oaks flourish in its hedges. Though its tendency to form mire does not make it ideal for residential purposes, its general flatness and its proximity to London encourage building 'development'.

Gault Clay The Vale of Holmesdale, which separates the Chalk Downs from the sandy line of hills farther south, is floored with a stiff blue clay called *gault* (a term the origin of which is unknown, but which may have been coined by the brickmakers). This bed extends to the sea near Folkestone, where its collapse has formed the Warren landslip. On the shore near by may be found nodules (lumps) of pyrites (iron sulphide), as well as crystals of selenite. The gault clay is rich in fossils, some of which retain a 'mother-of-pearl' lustre. It also contains nodules of a grey substance containing phosphates.

Clay-with-flints Here and there on the North Downs the rolling grassland is diversified by small beech-woods. These grow not on the chalk itself but on patches of a surface material known as Clay-with-Flints. This is a reddish clay, very sticky in texture, containing numerous flints which were left unaffected when the chalk in which they were once embedded was dissolved by the rain. It consists of impurities, similarly derived from the chalk, with perhaps the remnants of a layer of clay or sandy material which at one time covered the hills.

Boulder Clay Over much of the country north of the Thames are wider stretches of another surface deposit of clay. It shows no trace of arrangement in layers, but fills in the depressions in the rocks on which it lies. As its upper surface, though by no means flat, is less irregular than that which it covers, it has a 'smoothing' effect on the scenery. This *boulder clay*—so called because of the blocks of stone it contains—is regarded as evidence of glacial action (see page 201); a moving layer of ice, which formerly covered most of the British Isles, rubbed away fragments from the rocks, ground them to powder, and smeared them over the land.

Loam, Marl, and Brashy Soil Clay combines with other rocks to form valuable types of soil.

Loam is a mixture of clay and sand; *marl* of clay and limestone. A marly soil, such as that of the Central Plain of England, makes good agricultural land; so also does a *brashy* soil, one containing numerous fragments of limestone.

Anyone who has visited Cornwall must have been impressed by the curious conical hills, gleaming almost white in the sunshine, with their fanciful resemblance to the pyramids of Egypt. They are not of natural origin, but consist of the 'spoil' from the pits from which the china clay is obtained. Though so different in colour and appearance, they are analogous to the slag-heaps of the coalfields and other industrial areas.

Shale

Character   Clay consists of very fine grains, and these, like sand, silt and pebbles, can be welded into solid rock called *mudstone,* when it shows no trace of arrangement in layers. When it consists of thin layers, it is a *shale*, from an old English word meaning 'shell', or 'scale', and if the layers are sufficiently thin and sufficiently regular, so that they fancifully resemble the pages of a book, it is a *paper-shale*. The colour of shale is as variable as that of the clay from which it comes; and, again like the clay it may 'shade off' into other rocks. Sand-grains in the clay make it intermediate between sandstone and shale.

Oil-shale   Economically, the most important

type of shale is the *oil-shale*, in which the grains are, so to speak, soaked in combustible liquids, formed ages ago from the decaying remains of animals or plants. The oil is obtained by *distilling* the shales, exposing them to fierce heat in closed retorts. Sometimes, however, it takes fire spontaneously, making a shale-bed smoke and giving a hill-top a disquieting but misleading resemblance to a volcano.

The Blue Lias   To the geologist, however, the most interesting shales are those found near Lyme Regis in Dorset, near Whitby in Yorkshire, and elsewhere. These beds are so regular and so well marked as to earn the name of *lias* (quarryman's dialect for 'layers'). In the Lyme Regis area the soft beds of dark blue shale alternate with harder beds of bluish limestone, giving the cliffs a picturesque and unusual appearance. Viewed from the sea they look as if they had been ruled with a series of horizontal lines; they consist of parallel ledges about thirty centimetres thick and forty-five centimetres apart, making the cliffs seem as if they could be climbed by a natural step-ladder.

This impression, however, is false, for the cliffs are dangerous even to approach. Great masses of rock may break away from their face and crash to their foot, endangering anyone beside it. Similarly the top of the cliffs might easily fall away under the weight of anyone who ventures too near the edge. Lyme Regis is indeed notorious for its landslips, one of which involved acres of fallen cliff.

Blue Lias Fossils The lias of Lyme Regis is equally famous for its fossil wealth. Here were found many of the skeletons displayed in our museums, the remains of the great reptiles—the ichthyosaurus, plesiosaurus, and the like—which at one time dominated the earth. The amateur geologist is unlikely to make such 'finds', but a hunt along the beach is almost certain to reward him with a variety of smaller fossils. The most interesting of these are the ammonites, some of them made even more attractive by a coating of brassy-looking iron pyrites. (The smaller and prettier ammonites form a minor article of local commerce, being mounted in rings and sold as souvenirs.) Crystals of this mineral may also be found on the beach, though it does not occur in such quantities as would repay working.

The Parish Church of Lyme Regis contains an interesting stained-glass window dedicated to the memory of Mary Anning who systematically collected the fossils found in these cliffs.

Coal

Peat Much of the moorlands of Britain, and still more of those of Ireland, are soft and spongy to the tread. Dug into, they reveal a mass of a black or dark brown substance, with a pulpy feel and a fibrous texture. It consists of soil mingled with half-decayed vegetable material, the remains of reeds and sedges, of moss, or of heath. This material is *peat*, which country-folk dry to use as a low grade fuel; it burns only with a subdued heat but is practically smokeless.

Some peat-bogs are as much as fifteen metres deep, and show an unusual form of stratification, different types of vegetation being found in successive layers. The water which oozes from many moorland bogs is tinted a deep brown; this is due to the presence of *limonite* (bog iron ore, an hydrated oxide of iron), a mineral which formed under the influence of decaying vegetation. In Finland and Sweden, this ore has been commercially worked, and in Ireland the peat has been distilled to form producer-gas.

Lignite The process which turns vegetation into peat continues. Its next stage is 'brown coal' or lignite, which has been worked in the Weald and has heated the potteries of Bovey Tracey, Devonshire. Continued still further, it forms the *coal* on which our industries and the warmth of our homes depend. The various stages of the transformation produce different types of coal.

Cannel Coal *Cannel* ('candle') coal is so called because of the bright flame with which it burns. It does not soil the hands when touched, it breaks with a conchoidal fracture, and it has a misleading appearance of having been formed by the cooling-down of molten material. It is so hard and can so readily be polished that it has been used, like jet, for making ornaments.

Bituminous Coal *Bituminous* coal—wrongly named, as it contains no bitumen (pitch)—is commonly used for domestic heating. It consists of layers, some of which are alternately bright and dull, and is so soft as to soil whatever it touches. It contains a greater proportion of carbon than cannel coal.

Anthracite   *Anthracite* coal contains a still greater proportion of carbon. Like cannel coal, it breaks with a conchoidal fracture and does not soil the fingers, but it has a greater lustre and burns with an almost invisible flame. For domestic use it needs special stoves.

Graphite   Carbon in almost a pure state occurs naturally as *graphite* (from the Greek word for 'writing') sometimes incorrectly called 'black-lead' because of its appearance. Mixed with clay, it forms the 'lead' of lead-pencils, and was used in domestic polishes. It was at one time mined in Borrowdale, in the Lake District. (Strangely enough, carbon also occurs in a crystallized form. It is then no longer black and soft but transparent and extremely hard. The crystals of carbon are, in fact, *diamonds*.)

Coal-gas and Coke   Indispensable itself to our industries and our comfort, coal is also a raw material from which come many other valuable substances. Distilled, it generates *coal-gas,* itself a source of heat and light widely used in industry and in the home. The residue left from the process is *coke*—a material which does not occur in nature —another important fuel. The distillation does not only produce gas and coke; what was once thought to be a useless by-product, the coal-tar, yields a host of useful materials; among the coal-tar derivatives are numerous medicines and flavouring matters and the glorious colours of the aniline dyes.

Coal Fossils   The coal may bear the evidence of its origin in its fossils. Proverbially black itself, even blacker are the imprints of leaves and stems

Coal Fossils—Carboniferous ferns

on some of its surfaces. They are the remains of plants akin to ferns, club-mosses and horsetails which, however, grew not as lowly shrubs but as giant trees.

Coal Mining   The coal forms a number of thin *seams,* and alternating with these are layers of the soil, now transformed into clay, on which grew the vegetation from which the coal was formed. This is the fire-clay already referred to, and in it may often be found the fossilized roots of the coal-forming plants.

The needs of the Second World War demanded the working even of low-grade coals. So near the surface are some of these that they are reached by *open-cast mining*. In this process the overlying earth is excavated until the coal is exposed and is replaced more or less in its original order after the coal has been removed. Normally, however, coal is mined by means of pits sunk into the earth to reach the seams, along which galleries are dug. The work is not only laborious, but extremely dangerous, the roofs of the galleries having to be elaborately buttressed by means of steel pit-props and the coal in many pits continually exuding an explosive gas, *fire-damp*. In spite of all precautions accidents do occur; explosions of the gas cause underground fires and falls of the roof and produce a suffocating vapour, *choke-damp*. The industry has aroused a magnificent tradition of hardihood and devotion in its highly specialized workers.

Rock-salt

Cheshire Scenery A feature of the scenery of Cheshire is a number of shallow ponds, called *flashes,* filling gentle depressions in the ground. The houses in some of its towns have an odd look.

Some seem a little, or even markedly, askew, or stand upon concrete rafts; others have metal plates on their outer walls held in position by metal bars reaching from front to back or from side to side.

Rock-salt: Character These flashes, these braced-up houses, are the evidences of large-scale salt working. The great deposits of *rock-salt* which underlie the country are very different from the fine-grained white powder of our dinner tables. They consist of solid masses of gleaming material, white when pure but usually coloured brown, red, or yellow by iron or other impurities. Rock-salt is as soluble in water as table salt, and can as easily be recognized by its taste.

Salt Mining Rock-salt is rarely mined for with shafts and galleries like coal or any other mineral. It is commonly obtained less laboriously; water is simply pumped into and out of the earth, dissolving the salt during its flow. The brine thus formed is boiled and evaporated in huge tanks, its impurities— which themselves have industrial uses—being removed by ingenious processes involving the use of a 'twaddle' (a hydrometer to indicate the specific gravity of the fluid). The salt, which crystallizes out from the evaporating liquid, is packed by complicated machinery into cartons and packets for domestic use.

Ironstone

Impure ores of iron not only form irregular layers in the Lower Greensand. They occur also in large masses, as rusty-looking layers or nodules, in the coalfields, where they are called *clay ironstone,* and elsewhere. Some of them, as at Cleveland, Yorkshire, are oolitic (see page 91) with a grained structure, and have taken the place of beds of oolitic limestone. Others have a 'cone in cone' structure, consisting of conical masses one inside another. Clay ironstones, which mostly consist of carbonate of iron, form an important ore of that metal.

Meteorites

Unlike the other rocks which form part of the earth's crust, meteorites have fallen from beyond the atmosphere. In their rapid fall they become intensely heated, and their impact usually buries them in the ground. Fortunately, few of them are large, and almost all have dropped in the open country (one immense meteorite which fell in Siberia, early in the century, caused widespread damage to the forests). Some meteorites are 'stony', consisting of such silicates as olivine (see page 166), but most are formed of iron, alloyed with nickel and with small quantities of other metals.

ROCK FORMATIONS

Horizontal Rock-beds

The rocks of a district influence its scenery not only through such characteristics as their hardness and softness and chemical composition, the type of soil they form and the nature of the vegetation they foster, but also by their *structure*.

The rocks so far described are all sedimentary, having been formed through the hardening of layers of material deposited at the bottom or on the shores of seas, rivers and lakes. Such layers are naturally horizontal or gently sloping. In many places the strata still remain in the same position as that in which they were originally deposited, the hardened sediment having simply been upheaved bodily without being twisted or deformed. An example of horizontal bedding may be seen in the face of Malham Scar, Yorkshire.

If it were not for the weather, the horizontal rockbeds would form level plains or plateaux (elevated plains). Formed by accumulated rain, the rivers, however, wear away their surfaces, converting them into a number of hill-tops, all more or less at the same level. Such a 'dissected plateau' constitutes the hilly country a little inland from Lyme Regis.

The valley may cut through the surface layers and reveal rocks of an entirely different character below, the junctions of the various beds running horizontally along the valley-sides.

Folded Rock-beds

Most rock-beds, however, were twisted out of their original position during the process of up-heaval. Instead of being horizontal, or nearly so, they show a definite *dip* (slope). A few are vertical. Many are bent into gentle or abrupt curves.

Synclines A good example of gentle rock-folding is given in the London area. The chalk, and the rocks above and below it, underlie the city in a formation like a very flat saucer. Such a formation is called a *syncline*, from Greek words meaning 'slope together', the reference being to the dip—whether it be gentle or steep—of the formation's two 'limbs' (sides).

The dip in the North Downs, which is only about one degree, is towards the north. On the far side of the Weald is another series of chalk hills, the South Downs, and these are associated with similar rock-beds to those in the north—the chief difference being that here the sand is much thinner and does not form a lofty line of hills. These southerly beds, however, slope in the opposite direction from those in the north; their dip is towards the south.

Anticlines Separated from both North and South Downs by a tract of clay is a series of beds —mostly sand interspersed with clay and a little limestone—which forms the pleasant well-wooded country extending down the centre of the Weald from Ashdown Forest eastwards to the sea. The arrangement of these beds is the reverse of that below the Thames; they do not resemble a 'saucer' of rock but what might be called an arch, if the slope of its limbs were not so very gentle. Such an arrangement is called an *anticline*, from the Greek for 'slope against', because the two limbs dip away from one another.

Scenic Effects These structures, synclines and anticlines, or the two in alternation, are very common and have a marked effect on the scenery. They might be expected to produce corresponding slopes at the surface, making it rise in a hill above an anticline and dip in a valley under a syncline. Exceptionally—as in the Vosges—this effect may appear, but it is very unusual. The action of the weather on the folded rock-beds produces conditions which affect the scenery very differently.

When horizontal rock-beds are humped up to form an anticline, those at the top of the arch are stretched. Or rather, as rock is not elastic, they are cracked and broken and the fragments are torn apart. Rocks so loosened are easily destroyed by the weather. When the horizontal beds sink to form a syncline, on the other hand, those in the upper part of the trough are compressed. This renders them harder and more resistant.

Other things being equal, the weather consequently destroys the upper layers of the anticline more quickly than those of the syncline. It may level the humps of the anticline not merely to the troughs of the syncline, but even considerably lower. The synclines then form hills and the anticlines form valleys.

Had it not been for the weather, the rocks in the Wealden anticline would have formed a hill 760 metres high. Such a hill does not exist, however, for the weather destroyed it even as it slowly rose. It destroyed the chalk and the red sand completely, and the beds below the sand in part, all over the anticline, so that the highest points of this reach only a few hundred metres above sea-level.

The rocks in the Thames syncline—the London Clay and some beds of gravel—were, however, too soft to form a hill. The weather, later aided by the river and the tide, has worn them away until much of the area is only just above sea-level and even its loftiest points, formed by the gravels of Hampstead Heath, are comparatively low.

The highest ground in the whole region is formed by neither anticline nor syncline, but by the 'common limb' of both. This consists partly of chalk so resistant to the weather that it rises above the general level, as the North and South Downs, partly of sand so reinforced with layers of chert and ironstone that it rises even higher, as the Red Downs. Between these two resistant beds is a layer of gault clay so soft that the weather easily destroyed it, scooping out the narrow Vale of Holmesdale which separates the two lines of hills.

Beyond the Chilterns the chalk and its associated beds have been destroyed, exposing the softer beds below as the *Midland Plain*. Beyond the South Downs, the sea has swept away the land; we cannot observe what rock formations lie below the English Channel. But we can see that the alternations of syncline and anticline continue south-west of the Weald.

The Wealden anticline is followed by a small syncline, and this by a smaller anticline—the Portsdown anticline, near Portsmouth. Its farther limb slopes very gently southwards as though to form another syncline, but instead of doing so it bends almost perpendicularly at its lowest point, rising almost vertically to form the central ridge of the Isle of Wight. Such a formation is called a hogback or *monocline* ('single slope'). The beds then again bend perpendicularly, resuming their gradual southward dip, though at a higher level than before. Here the chalk has been largely destroyed by the weather, but a broad outlier remains to cap St. Catherine's Down, the highest ground in the island.

Vertical Rock-beds

An outcrop in which the rock-beds are almost vertical has noteworthy scenic effects. The ordinary inclined bed produces an asymmetrical hill, steeper down its escarpment—the slope which ends the dip—than down its dip-slope; compare, for example, the gentle rise on the London side of Box Hill and Leith Hill with the steep descents which these hills present to the south. The hill produced by vertical beds is much more symmetrical, both its sides having much the same slope.

Inland it forms a narrow ridge rising abruptly above the lower ground on either side; an example is the Hog's Back stretching from Guildford nearly to Farnham, Surrey. On the coast it produces a steep cliff towering above the sea. At their eastern extremity, the vertical rocks in the Isle of Wight produce Culver Cliff; at their western extremity, the cliffs of Tennyson Down, the narrow promontory overlooking Alum Bay, and the detached islets of the Needles. Farther west, the formation continues as a double line of hard vertical beds, the chalk which forms the cape north of Swanage and the cliffs beyond the Durdle Door, and the limestone which forms the cape south of Swanage and the cliffs near Lulworth Cove.

Contorted Rock-beds

The rocks may be bent into more complicated patterns than the gentle undulations of south-eastern England and the monocline of the Isle of Wight. They may form *contorted strata* consisting only of small-scale zig-zag folds, like those exposed in a roadside cutting on the highway between Newport and Gloucester and in the cliffs at Stair Hole, near Lulworth Cove; or they may involve crumplings of astounding complexity extending through whole mountain ranges.

The origin of these formations is illustrated by placing a pile of cloth or paper below a heavy weight and then pressing its sides towards one another. The pressure forces the layers of cloth or paper into crumplings very like those seen in the Scottish Highlands and the Alps.

Earth-movements It is difficult to imagine a great thickness of solid rocks being forced, no matter how powerful the pressure, into such con-tortions. Certainly the thrust cannot have been sudden and violent, or the rocks—very few of which are plastic—would have been not folded but smashed. Only a very steady pressure, con-tinued over a very long period, could produce such a result. The cause of such earth movements is eagerly debated by the geologists, but their theories, though interesting, are beyond our present scope. Even more interesting is the evidence they find of similar thrusts in operation at the present day!

Mountain Building   One such disturbance upheaved the earth's crust into a series of mountains.
Its centre was in Asia, where it produced the
Himalayas; so far away as central Europe it was
powerful enough to form the Alps. In England,
where it was comparatively feeble, it merely
caused the undulations in the rock-beds from the
Chilterns to the Isle of Wight.

This was the most recent of several periods of
'mountain-building' of which geologists find
evidence. The mountains of Wales, the Lake
District, and Scotland were produced by earlier
disturbances. During the long periods which
separated these upheavals, these mountains were
largely destroyed by the less spectacular but
nevertheless destructive effects of the weather.

Overfolds   Only vestiges remain, comparatively
speaking, of the great mountain masses produced
by the upheavals. They none the less bear evidence
of far-reaching disturbances. The rocks are liter-
ally crumpled. They are not merely bent but are
overturned and piled into what look like flattened
loops, called *overfolds,* so that in some places a
series of rock-beds are in the exact reverse of their
customary order. Movements so violent, of
course, produced remarkable changes (see pages
154–5) in the character of the rocks affected.

Faults

Cracks in the Earth's Crust   Subject as they
are to such tremendous forces, and also to less

powerful but more sudden strains, it is not surprising that the rocks should be traversed by a large number of cracks. These have displaced the strata, the beds on their two sides being obviously out of alignment. Some displacements are matters only of a few centimetres; others cleave through rock-beds and separate the halves by distances to be measured in kilometres.

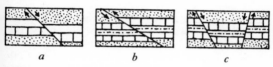

a   *b*   *c*

Faults—Normal (a); *thrust or reverse* (b); *trough* (c)

Such a crack in the earth's crust is known as a *fault*. The slope of the *fault-plane* is not its dip but its *hade,* and the vertical distances separating corresponding strata on its two sides is its *throw ;* the sides are called the *upthrow* and the *downthrow* respectively, depending on which of the severed beds are higher or lower. These terms, as might be expected from their terseness, were coined not in the study but in the pit, the faults having forced themselves on the attention of the mining engineer and the miner before the geologist began to investigate them.

Normal Faults   The *normal fault,* the type most frequently encountered, one that 'hades to the down-throw'; it was produced by a force which pulled the rocks apart, and looks as if the one half had slid bodily down the fault-plane, coming to rest some distance below its original position. The extra work it gives the miner is obvious; a seam of coal or ore ends abruptly, and its continuation has to be sought for on a different level. Moreover, a trial-boring which happened to pass between the severed ends would totally fail to disclose the existence of the seam.

Thrust Faults The less common *thrust-fault* is one which 'hades to the upthrow' and is the result of sideways pressure, which snapped the rock-beds and made their broken ends overlap. Here again the miner would find his seam end abruptly, and have to seek for it on a different level. A trial-boring made through the overlap would give a very different impression from that made between the severed beds in a normal fault; but it would be no less misleading and no less fraught with unpleasant consequences. The one might mean that a valuable seam was overlooked and unworked; the other, by reaching the same seam twice at different levels, would suggest that two seams existed and might cause needless expenditure of capital and effort only resulting in disappointment.

Overthrusts A variant of the thrust-fault is the *overthrust*. Its fault-plane hades are almost horizontal, a thin section of the earth's crust having been driven by a powerful sideways thrust bodily over another. It may result from an overfold (see page 74) so violent as to snap the contorted rocks.

Tear-faults   Earth movements may be not so much up or down as sideways, resulting in a *tear-fault*. The destructive San Francisco earthquake of 1906 involved a sideways movement of a few metres. Such a movement would sever a seam of coal or ore, though its continuance would have to be sought on the same level.

Slickensides   The rocks along the fault-plane may be scratched and grooved as their faces scraped together. They may be polished by the friction, producing a gleaming surface known as a *slickenside*. They may be crushed along a *fault-zone* of some thickness into a natural rubble called a *fault-breccia*. Molten or dissolved material welling up through the shattered rocks may produce a vein of valuable minerals.

Scenic Effects   When newly formed, a fault may cause a visible inequality, like a natural step, in the ground, but this is soon destroyed by the weather. A more permanent effect on the scenery is produced when the fault with a large throw brings side by side at the surface two rocks differing greatly in hardness.

A spectacular example of the effect of faulting is given by the Scottish Highlands. Their sudden rise from the Lowlands is caused by a series of faults—some with throws of hundreds of metres—along a line from Stonehaven, Kincardineshire, to Dumbarton. Another series of faults, from near Dunbar to Girvan, Highland, causes the hills of the Border country to rise abruptly south of the Lowlands.

Even when a fault produces no obvious effect

on the scenery, its existence may be deduced by a study of the outcrops. It may interrupt the rock-beds at the surface and separate the severed ends. By cutting a rock-bed short below the ground-level, it may prevent it from coming to the surface at all. Given certain relations between the dip of the strata and the hade of the fault-plane, it may bring the same bed to the surface twice.

The fault, however, may not reach the surface. It may be covered merely by thin layers of soil, silt or boulder-clay; or it may cease abruptly some distance underground, other layers having accumulated and hardened into solid rock since it was formed. Then the geologist will have to infer its existence and estimate its throw by an elaborate process of deduction; the value of this work to the mining-engineer is obvious.

Normal faults occur almost everywhere. Thrust-faults and overthrusts affect the very rocks that show violent contortions, and have been produced by associated forces. All three are found in bewildering profusion in that complex geological region, the Scottish Highlands.

Step and Trough Faults Faults may occur simply, or intermingle to form a complicated pattern. A large fault may 'fade out' at each end, or branch into smaller faults extending in different directions. A series of faults, all throwing in the same direction, may run parallel; these are called *step-faults*. Two normal faults may have their downthrows facing inwards, so that the rock-beds between them are lower than those outside; this forms a *trough-fault*.

The trough-fault indicates that the rocks have been rent apart by powerful forces, leaving a triangular gap into which a wedge-shaped mass of material has sunk. One such rending, accompanied by a sideways displacement of about ninety-six kilometres, produced a groove in the earth, in one place 300 metres deep. This is the Great Glen of Scotland, the northern boundary of the Highlands; it includes the largest body of water in the British Isles. The depths of this are said to harbour a strange creature unknown to science—it might perhaps be regarded as a 'living fossil'—the Loch Ness Monster.

Unconformities

Certain rock-beds form a regular series, mostly lying consistently in the same order wherever they may be. In rare instances, however, there is a gap in the sequence; a bed (or beds) is missing where, by all analogy, it ought to occur. Such a formation is called, somewhat clumsily, a *non-sequence*.

In many places a number of different rock-beds —whether they are horizontal, sloping, vertical, or contorted—run parallel for a very long distance, for a total thickness of hundreds of metres. Elsewhere adjacent beds dip at different angles. The difference may be very slight, so that it is only evident over long distances, or it may be great. Horizontal or gently sloping strata may visibly lie on the ends of vertical or steeply sloping beds, or on the sheared-off tops of highly contorted folds. Such a difference in the dip of adjacent beds forms an angular *unconformity*.

A bed of stratified rocks gives evidence of one change of level, an elevation of a former sea-floor to produce dry land. An unconformity indicates *two* such movements, separated by the depression of a land-surface below the sea. If it is accompanied by a difference in dip, it testifies to more complicated earth movements.

The gap in a sequence of rock-beds shows that the region where it occurs formed dry land during the period when the missing layers were being deposited elsewhere. Hence it must have risen after the rock-beds below the gap were formed; and while it remained above sea-level the weather wore irregularities in its surface. Then it must have sunk below the sea for the beds above the gap to accumulate. Finally, it must have risen a second time for them to become dry land.

If these movements were simple changes of level, both series of rock-beds—below and above the gap—would remain parallel. A difference in dip—producing an unconformity—shows that the lower beds were not only raised but were tilted or contorted. The weather planed away their tops to form a surface, more or less level; then the beds sank and then emerged for the second time.

THE RECORD OF THE ROCKS

Bygone Seas The bygone seas and lakes in which the various sediments were deposited no more formed a continuous body of water than do those of today. They were bounded by shore-lines smooth or jagged; they varied in depth; they had their waves and currents and tides; they were affected by local peculiarities and climatic conditions. The sediments .which accumulated on their floors were correspondingly varied, and the variations have their effects on the rocks formed by the hardened sediments.

A bed of rock may 'thin out' at its edges, or be interrupted by some bygone island or cape. Another, though plainly continuous over a wide area, may vary greatly in composition from place to place. A limestone may be increasingly affected by muddy impurities until it becomes a shale; a shale may coarsen into a sandstone and so into a conglomerate. One rock-bed may overlap another, the two having been formed on a gently sloping beach while this was slowly sinking into, or rising out of, the sea.

Prehistoric Geography An interesting branch of geology, sometimes given the impressive name of *Palaeogeography* (*Palaeo* coming from a Greek word meaning 'ancient') attempts to reconstruct, by study of the modern rock-beds, the obliterated

outlines of lands and seas which have long
vanished from the face of the earth. This is a
matter for the systematic geologist; but we can
follow the principles which guide him in his work.
We can see the significance of ripple-marks on a
sandstone, of a hardened ooze similar to that of
the modern ocean-floor, or of the abundance in a
rock-bed of fossil sea-creatures or land plants.

In itself geography—the layout of sea and land
—is not uninteresting. But when the land and sea
are regarded as the homes of living creatures
greater interest arises, and this is as true of the
distant past as it is today. Who but a specialist
would care to map, through millions of years, the
changes of the coast-line? Yet the interest of most
of us is aroused when we find buried in the rocks
the unmistakable remains of extinct animals or
plants. The bygone landscapes 'come alive' in our
minds when we imagine them as peopled with
living things.

Fossils These remains are known as *fossils,*
from a Latin word meaning 'things dug up'. They
are so plentiful that nearly all the bedded rocks
contain them, and some almost consist of them.
Their real nature was once doubted—they were
vaguely regarded, for example, as 'sports of
nature'—but they exist in such multitudes, and
some of them in such perfect detail, that it is
beyond dispute.

The fossils vary greatly from rock-bed to rock-
bed, so much so that each bed can be identified
by its characteristic fossils. The fact was discovered by the 'father of English geology',

William Smith, over a hundred years ago, and later study has amply confirmed his discovery. It is, indeed, the foundation of the whole science of *Palaeontology* ('knowledge of ancient life').

Geological Maps   If the reader glances at the captions to a geological map, he will see that few of them—the chalk and the Old Red Sandstones are exceptions—indicate the material of which the rocks are formed. They mostly consist of unfamiliar and difficult names, the meaning of which is far from obvious. They form part of the system by which the rocks are classified not by the material which forms them but by the fossils which they contain. These change in character according to the comparative age of the rocks, an age which is deduced from the order of the rockbeds.

Order of Superposition   When the sediment slowly accumulates at the bottom of the sea, its lower beds are obviously deposited before the upper ones and are therefore older. Similarly, of two parallel beds of rock—unless they have been bodily overturned, a process which is usually easy to trace—the lower bed will be older than that above it. In unconformities, again, the upper series of beds are more recent than those on which they rest. The sequence indicated both by the arrangement of the rocks and by the fossils which they include is confirmed by whatever radioactive minerals they contain—these convert themselves into lead at a measurable rate.

Classifying Rocks This classification of rocks according to age is largely based on the general character of their fossils. It divides them into about a dozen main groupings, called either *systems* or *periods* depending on whether interest centres on the rocks themselves or on the respective times during which they were formed. The different groups have complicated subdivisions, and the periods themselves are grouped into five *eras*.

The method of naming these groups is no more logical than are our spelling, or the notation of our piano-music. Like these, however, it works reasonably well, and it is no more likely to be altered than they are. Illogical as the terms may seem, they are a great aid to understanding the fossils. They help us to 'fix' the various extinct creatures in space and time, and to grasp their relation to one another.

The Geological Eras The geologists at first distinguished three great eras, and naturally named them, according to their order, as *primary, secondary* and *tertiary*. Later they added the *quaternary,* though this includes so few beds, and indicates so short an era of formation, that H. G. Wells, in *The Outline of History,* said that it was like separating the last page of a three-volume book and calling it volume four! They alternatively named them according to the type of fossils they contain, using—as usual—words derived from the Greek. This alternative method is most used, because 'primary' and 'secondary' have other geological meanings. Moreover, further investigation showed that there was a great thickness of rock-beds far older than the so-called 'primary'.

The Age of Ancient Life

Archaean Rocks   This oldest era is the *Archaean* (from the Greek word for 'ancient' or 'beginning'). As it contains no *definite* fossils, but only obscure traces, possibly of living things, and minerals which may have been formed from living matter, it is also known as the *Azoic* (Greek for 'without life') or *Eozoic* ('dawn of life'). Unlike the other eras, it cannot be clearly divided into periods. It is called *Pre-Cambrian*, because it is earlier than the first recognized period. Its rocks form much of the Scottish Highlands, and also occur in Dyfed, in Gwenydd and in a few places in England. They consist of crystalline rocks completely altered from their original character, together with some intensely hard sedimentary rocks. The period during which they were formed is estimated to amount to a hundred million years, yet the rocks from whose fragments the oldest pre-Cambrian sedimentary beds were formed were obviously older still.

Palaeozoic Rocks   The second of the great eras, formerly known as the Primary, is called the *Palaeozoic* (Greek 'Ancient Life'), because almost all its fossils are so very different from the plants and animals of today. Its rocks are estimated to have taken six hundred million years to form. It is divided into Lower and Upper, each of the three systems which form the Lower Palaeozoic being named after the region in North and Central Wales where it was first studied.

Cambrian Rocks   The oldest beds containing *definite* traces of living things are called the *Cambrian* system, after the ancient name for Wales. They occur not only in North Wales—in Merionethshire and along the Menai Straits—but in Salop and the Malvern Hills, in the Northwest Highland and in the Isle of Man. They consist of very hard sandstones and shales and slates, with some grits and conglomerates. Their fossils are of lowly animals and plants; they include markings left by moving seaweeds and crawling worms, and the remains of shellfish and the curious animals known as trilobites (see page 115).

Ordovician Rocks   The *Ordovician* Rocks (called after the Ordovices, an ancient Welsh Tribe) are found in North, Central and South Wales, in Salop, in the Lake District, in the South of Scotland and in eastern Ireland. They consist of dark-coloured shales, grits and sandstones, with some limestones. Their fossils are somewhat similar to those of the Cambrian System but also include the earliest fish—of types very different from those of today. This period was one of great volcanic activity, so that many igneous rocks (see page 152) are interbedded among the shales, sandstones and grits.

Silurian Rocks The *Silurian* Rocks (called after another Welsh tribe, the Silures) occur along the Marches of North Wales, and in Scotland just north of the Border. These also consist of shales, slates, sandstones and limestones; their fossils, though again of lowly types, include the earliest land animals and the forerunners of the ammonites (see page 120). The rocks of all these three systems are generally similar, as are the fossils they contain; indeed, controversy as to the dividing line between Cambrian and Silurian was only settled by recognizing the Ordovician between them.

Devonian Rocks The oldest system of the Upper Palaeozoic includes two separate groups of beds, very different in character but of roughly the same age. The *Devonian* Rocks occur chiefly in the north and south of Devon, forming the two arms of a great syncline whose centre consists of more recent beds. Those in the northern area extend from Morte Point near Ilfracombe right across and beyond Exmoor—Dartmoor, as will be seen later, consists of rocks very different in character—and on into North Somerset. The southern beds might equally be named after Cornwall, for they form a coastal strip nearly from Penzance to Paignton. The Devonian rocks are mostly sandstones, shales, grits and slates; there are also thick beds of limestone near Plymouth and Torquay. Except for the limestone, which largely consists of corals with some lamp-shells and trilobites, they contain few fossils.

Old Red Sandstone   The Devonian rocks were formed at the bottom of the sea; the *Old Red Sandstone,* though roughly of the same age, was deposited in lakes and coastal inlets. The origin of the name is obvious; the rock consists of sandstones and conglomerates so rich in iron as to give their own striking colour to the soil of Powys and Herefordshire, of the Cheviots, of the Midlands Valley of Scotland (south of Highland), of the John O'Groats region and Orkney. The fossil fish which abound in the Scottish beds are admirably described in a classic of geological literature, *The Old Red Sandstone.* (Its author, Hugh Miller, was a stone-mason who became one of Scotland's pioneer geologists.)

Carboniferous Rocks   The next series has three subdivisions. The *Lower Carboniferous* ('coal-bearing') consists of the Mountain Limestone. The Millstone Grit and the Coal Measures themselves together form the *Upper Carboniferous.*

Mountain Limestone   The *Mountain Limestone* produces the high ground of the Mendips; of Derbyshire; of part of the Pennine Chain; of Scotland south-west of the Forth Bridge; and of the greater part of Ireland. It is very rich in corals, sea-lilies, lamp-shells and other fossils.

Millstone Grit   The *Millstone Grit* of the Pennine region consists of sandstones, shales, and the very coarse sandstones, admirably suited for the grinding of corn, known as grits; in South

Wales it is more shaly. It was formed in shallower water than was the limestone and its fossils are therefore very different. They include two-shelled molluscs and the earlier varieties of *ammonite*.

Coal Measures In the *Coal Measures* the actual coal-seams do not form so great a thickness of rocks as the shales, sandstones and clays associated with them. They form much of our 'Black Country': the coalfields of South Wales, Wrexham, the Midlands, Northumberland, and Central Scotland. The coal consists of fossil vegetation, and the measures also contain fossil insects, molluscs, fish and amphibians.

Permian Rocks The topmost Palaeozoic bed, the *Permian*, is called after the Perm district of Russia, where similar rocks are well developed. It includes the strip of magnesian limestone (dolomite) which runs southwards from the Durham coast, together with some of the red marls and conglomerates which form part of the New Red Sandstone.

The Age of Middle Life

Mesozoic Rocks The greater part of the New Red Sandstone belongs however not to the Upper Palaeozoic but to the lower part of the *Mesozoic*. This name comes from the Greek for 'Middle Life' because its fossils are those of plants and animals widely different from those we know but not so completely different as their Palaeozoic predecessors. The Mesozoic is emphatically the 'Age of Reptiles', because these were the dominant form of life throughout the era. Its rocks took nearly two hundred million years to form.

Triassic Rocks   Because in Germany, where it is especially well developed, it has three well-marked divisions, the lowest Mesozoic system is called the *Triassic*. In England, however, it has only two divisions, the *Bunter Sandstone* (from the German for 'Bright-coloured') below the *Keuper Marl*. Both are a brighter red than the Old Red Sandstone and consequently form a lighter-coloured soil. These rocks constitute the north-western part of the Midlands; they include the salt-beds of Cheshire and large deposits of gypsum (calcium sulphate). They are almost completely unfossiliferous, and were formed in desert conditions, diversified here and there by dried salt-lakes.

Jurassic Rocks   In great contrast to these are the rocks which form the second Mesozoic system, called the *Jurassic* after the Jura Mountains on the borders of France and Switzerland. These also have two subdivisions, both very different from one another and from the Triassic beds below, and both very rich in fossils.

The Lias   The Lower Jurassic beds are also called the *Lias* (quarrymen's dialect for 'layers') because their stratification is so unusually obvious. They form the blue cliffs, consisting of alternate beds of limestone and shale, at Lyme Regis and Whitby, and include the whole strip of low-lying country between.

The Oolite   The Upper Jurassic contains many beds of limestone, of the type called the *Oolite* because its grains are so closely packed together as to suggest the roe of a fish. It forms the Cotswold Hills and their continuation; the limestone ridge which crosses England, east of the Lias and the Trias, from Dorset to Yorkshire. This system also includes the clays above the Oolite, which form flat marshy country, that part of the Midland plain which borders on the Chilterns as well as the Fens.

The Jurassic beds are very fossiliferous. It is this system which yields the remains of such gigantic reptiles, marine or land-dwelling, as plesiosaurs, ichthyosaurs, dinosaurs (see page 128), as well as more familiar types like turtles and crocodiles. It abounds in smaller fossils, for some of the limestones consist almost completely of shells or corals, and contain countless ammonites and belemnites (see pages 123–4).

Cretaceous Rocks   The most recent Mesozoic system is called the *Cretaceous*. Though this term comes from the Latin word for 'chalk' (compare the French *craie* and the German *Kreide*), it is not restricted to the chalk itself. Indeed, the *Lower* Cretaceous is well below the chalk, and comprises the *Wealden Beds,* of varied clays and sandstones; the Lower Greensand, a thick mass of red and yellow sand traversed by irregular layers of Clay Ironstone; and the lower part of the Blue Gault Clay. The *Upper* Cretaceous includes the upper part of the Gault, a much thinner sandy layer, the Upper Greensand, and the whole of the chalk.

The Cretaceous fossils, though generally similar to those of the Jurassic, show noteworthy differences. Among them are the last of the monstrous reptiles, and ammonites of unusual shapes instead of the usual closely packed spirals. The chalk fossils include sea-urchins, sponges, and shellfish, fish, birds of a primitive type and the remains of evergreen trees.

The Age of Recent Life

Tertiary Rocks Perhaps because it has no confusing ambiguity of meaning, the term *Tertiary* is often used for the fourth geological era. Its alternative name, the *Caenozoic* ('recent life') refers to the growing resemblance of its fossils to modern plants and animals. Its reptiles are not fantastic monsters, but smaller and more ordinary; they are unimportant compared to its birds and beasts. Its dominant type has led to its being called the 'Age of Mammals'; a very brief period, comparatively speaking, the whole era being estimated to have taken only sixty million years to form.

Its rocks, which in this country are separated from the Cretaceous by an unconformity, are as different as its fossils from their predecessors. Few of them, indeed, are hard enough for the man in the street to think of them as 'rocks'; they are mostly sands and clays. They are small both in thickness and in extent, much of the British Isles having consisted of land throughout the era.

The different systems which form the Caenozoic are distinguished by the proportion, in their fossil shells, of types similar to those now living. Their names, appropriately enough, are based on the Greek word for 'recent'.

Eocene Rocks The *Eocene* ('dawn of recent') includes the London Clay. Above and below the clay are beds of sand and gravel which form the heathy pinewood country of West Surrey and cap such hills in the London region as Harrow and Hampstead. Their hardest rocks are the Hertfordshire Puddingstone and the odd sandstone blocks known as Greywethers and Sarsen Stones.

Oligocene Rocks The *Oligocene* ('few recent') is practically limited to the Hampshire coast and the northern part of the Isle of Wight. The *Miocene* ('middle recent') does not occur in these islands, all of which were above the sea when it was formed.

Pliocene Rocks The *Pliocene* ('more recent') forms the shelly sand-beds of East Anglia, known locally as *crag*. Odd 'pipes' (see page 40) of sand in the chalk of the North Downs, and perhaps some of the Clay-with-Flints on its surface, also belong to this period.

Short as the Caenozoic was, it witnessed remarkable changes of climate. Tropical conditions developed in the Eocene, but these passed away, to be followed towards the end of the era by a growing tendency to cold.

The Present Geological Age

The Quaternary   The fifth geological age—the one still in progress—is known simply as the Quaternary. Its rocks are very similar to those of the Tertiary and its earlier fossils not strikingly different. The thinness of its beds indicates that the time it took to form is to be measured not in hundreds of millions of years; it is merely about a million. It includes only two systems.

Pleistocene Rocks   The *Pleistocene* ('most recent'), in which all the sea-shells are of modern type, includes the Boulder Clay which covers much of our islands north of the Thames. This was formed during a period of intense cold, the *Great Ice Age,* when the polar ice-cap extended to the River Thames; when the animals which roamed Europe—woolly mammoth and rhinoceros, cave-bear, reindeer and bison—were protected by coats of shaggy fur; when strange creatures, in certain respects resembling apes and in others resembling man, left not only their bones but their clumsy tools and weapons buried in the glacial drift.

Recent Rocks The second Quaternary system, the *Recent* (sometimes also called *Holocene*) includes the beds now in process of forming—the peat-bogs, the silts deposited by the rivers and the sand and pebbles washed ashore by the sea—and so brings the record of the rocks up to date.

The Age of Man It is easy to see the force of H. G. Wells' contention that these beds are too insignificant to be regarded as a separate era. Surely they merely represent the topmost system —or at most two systems—of the Caenozoic! Is it not our extreme pride that makes us give it so special a treatment merely because its fossils include human remains?

There is, however, good reason to treat the Quaternary, brief though it may be, as a separate era. Man appears in its beds not merely as a source of a growing complexity of fossils. From small, insignificant, and doubtful beginnings he becomes in himself a geological agent of increasing importance, comparable with the forces of nature.

Like these forces, he can both destroy and construct. He quarries the mountains, excavates the minerals, produces widespread soil-erosion by reckless forest-felling and intensive cropping; he destroys plant and animal species; he threatens to blast wide areas with his atomic bombs. Yet he reclaims the fens and the marshes, confines the flooding rivers into narrow channels, wins new land from the sea, grows new forests and irrigates the deserts; by selective breeding he transforms animal and plant; he converts waste country into pasture and cornfield and a human home.

The use—or misuse—of the latest human achievements is likely to transform the world, whether for good or for evil, to an extent which we cannot yet foresee.

The period which produced a strange biological species able to release the energy of the atom may be short. But it is certainly not unreasonable to regard it as a distinct geological era.

FOSSILS

Definition The term *fossil* ('thing dug up') at
one time meant anything excavated from the
earth. It is now applied only to the remains—or
traces—of 'things' which had life. It covers not
only their actual parts but the impressions which
they made. The trails of crawling jellyfish, the
holes bored by shipworms, the faint marks made
by the fronds of waving seaweed, may be per-
petuated in the rocks. Then they become fossils
as much as the bones of an ichthyosaurus preserved in the Blue Lias of Lyme Regis.

Conditions of Fossilization Though many
rocks teem with fossils, these represent but a
small proportion of the earth's inhabitants. Only
those creatures whose bodies include hard resistant material are likely to be fossilized. Their
soft parts are eaten, or decay, very quickly, and
rarely do they become fossils. Hence most fossils
consist only of hard substances; tree-trunks or
roots, animal shells, bones, or teeth. The hard
parts themselves are likely to be preserved only if,
sinking to the bottom of the sea, they are covered
with a sediment which hardens into rock and
later rises above sea-level. For this reason fossils
of the inhabitants of seas and swamps are much
more plentiful than those of land-dwellers.

Many fossils go undiscovered. The beds which
contain them remain inaccessibly on the sea-floor.

Even those raised to form dry land are soon destroyed, with all the fossils they contain, by the weather, or by the large-scale processes of mechanical quarrying. A labourer working with pick and shovel may spot some interesting object and put it aside; but the bulldozer rips away the rock-beds indiscriminately, regardless of their contents.

Methods of Fossilization Flies entrapped in amber, mammoths entombed in the arctic snows, are preserved in their entirety. Actual bones or shells may also be preserved, though bones embedded in the more recent sands or clays may fall to pieces, unless they are very skilfully handled on being removed. (Instructions for dealing with such fragile objects are published by the Natural History Museum, South Kensington.)

Fossil Moulds Though some of the hardest bones and shells decay, they may last long enough for the silt in which they are embedded to harden into rock. Then, even after their decay, they leave their impression in the rock, a sort of natural *mould*. This need not necessarily be an impression only of their exterior; a sediment may seep into, and harden inside, a sea-shell, producing a mould of its interior surface.

Fossil Casts After the object decays, the sediment may similarly ooze in to fill the vacant space it has left. This sediment also hardens; but it may differ somewhat in colour or texture from the hardened sediments around it. It may not blend with these but become a tiny nodule of rock, filling the natural 'mould' even more tightly than a kernel fills its nutshell. It then becomes an exact replica, so far as the surface goes, of the original object; it is a natural *cast*.

These moulds and casts, most of which occur in sandstone, are fossils, so to speak, only superficially. They have none of the interior structure of the original object. Broken, they reveal themselves as consisting only of solid rock—differing a little, perhaps, from the beds in which they lie, but of no special interest. So far as the surface goes, however, they may be perfect; they may preserve even the most delicate details. A cast or mould of a shell's interior surface, for example, may show the attachments of the muscles, and thus enable their structure to be deduced.

Molecular Replacement   Very exceptionally a fossil may be preserved not superficially but in its entirety. A particle of mineral matter—calcite, silica, or iron-ore—may replace every particle of plant or animal matter. Thus a fossil tree-trunk may consist not of wood but of silica, the mineral which forms sand, quartz, or flint. In common speech, fossils are sometimes said to be 'petrified' (turned into stone) but the term is accurate only when applied to animals and plants thus preserved by *molecular replacement*.

Occurrence of Fossils   Fossils are found in all the geological systems, from the Cambrian to the present day. They cover almost every type of animal and plant there is—and many that do not

99

now exist, their fellows having long ago become extinct. For the reasons already explained, they do *not* include certain types—those lowly creatures, for example, whose bodies contain no hard parts to be preserved. The former existence of some very interesting animals is known only from isolated specimens.

In everyday life we are most interested in the larger animals—those whose bodies are arranged on the same general plan as our own: the beasts, the birds, and to a lesser degree the reptiles and the fish. The more common fossils are, however, very different from these: the remains of such small sea-living creatures as shellfish, sea-urchins, corals and sponges. The amateur *palaeontologist* ('student of ancient life') will find thousands of these for every bone of bird or beast that he discovers.

Microscopic Fossils

The simplest living creatures are the tiny blobs of moving jelly, called *amoebae*, which the microscope discloses in pond water. Each consists of an isolated *cell,* somewhat resembling the units which co-operate in millions to build up plant and animal bodies. The earliest living things are supposed to have been somewhat of the same type; as they possessed no hard parts, however, they naturally never became fossils.

Some one-celled creatures, in spite of their simple structure, produce microscopic skeletons or shells. Those in which these consist of calcite are called the *Foraminifera*: multitudes of their shells help to form the Atlantic ooze and build the limestone and the chalk. Those in which the skeletons consist of silica are called the *Radiolaria*; their remains, also found in the ocean ooze, occur in some specimens of chert. Plants equally small and equally simple are called *diatoms*; their hard parts accumulate to form an abrasive material known commercially as 'Tripoli powder'.

Radiolaria (magnified)

Beautiful as these tiny fossils are, it takes a microscope to detect them, and hence they are beyond the scope of the outdoor observer. As, however, they help to form the rocks which produce the scenery, they are worthy of his attention.

Fossil Plants

The Earliest Plants The smallest fossil plants are the *diatoms*. The largest are whole tree-trunks converted into stone. There is evidence that life began in the sea, and the earliest plants, which left doubtful traces in the Cambrian rocks and definite remains in the Silurian, were seaweeds.

The earliest land-plants, of which there are many fossils in the Old Red Sandstone, mostly resembled ferns, club-mosses and horsetails. They grew, however, not as lowly shrubs but as great trees. There were as yet no flowering plants, and the forests of the time must have been sombre and monotonous.

Coal-forming Plants These forests, though with a somewhat greater variety of flowerless trees, flourished even more abundantly in the Carboniferous period. Growing on mud-flats and in swamps on a slowly sinking land surface, they were buried beneath the silt brought down by the encroaching waters. Compressed by the growing weight above them, they suffered chemical changes —they were 'mummified'; hardened and blackened, they became coal. Imprints of stems and leaves and bark may sometimes be detected on flat surfaces of coal—or on the shale which accompanies it—and fossil roots spread in the deposits of fire-clay below the coal-seams.

Similar plants lived throughout the rest of the Palaeozoic and well into the Mesozoic. In the Jurassic appear the ancestors of our fir-trees (and of the aptly named 'Monkey Puzzle'). The prevailing character of the Lower Cretaceous forests was also of this flowerless type.

Flowering Plants Fossils of flowering plants, though no doubt these appeared earlier, first became abundant in the Chalk. Those of the earlier Tertiary included types characteristic of a tropical climate. These died out as conditions became more temperate; and, in spite of such vicissitudes as the Great Ice Age, the woods of Britain gained their present aspect.

Fossil Forests The most abundant fossil-plants are those which form the coal. The most remarkable are the Fossil Forests. They indicate at least a double motion of the shore. The trees grew above sea-level. In order to become 'petrified' they must have sunk below the waves; and now they are again on dry land. They may be seen in several places round our coasts: in Swansea Bay, Barnstaple Bay, and near Stornoway in the Island of Lewis. St Michael's Mount, now an island surrounded by sand, bears a Cornish name which means 'the hoar rock in the midst of the woods'.

Very different from these is the Fossil Forest on the edge of the cliff near Lulworth Cove, Dorset. It is far older than the submerged forests on the beaches, and forms part of an extensive bed of limestone, of Mesozoic Age, in which many fossil trees have been found.

Fossil Sponges

Modern Sponges The simplest many-celled animals are the sponges. Their bodies are much less unified than those of other animals; it would be hard to decide whether a sponge is one individual or a colony! Each sponge has a skeleton perforated by many small holes, through which it absorbs water; having extracted its nourishment from this, it expels the water through a single large opening. Our ordinary domestic sponges are formed of these skeletons, most of which consist of a horny substance called *chitin*.

Sponge Spicules Though the fossil sponges are skeletons, they consist not of chitin but—as do some modern sponges—of calcite or silica (or

very infrequently of iron ore). They may remain intact, somewhat resembling wine-glasses pierced, network fashion, by numerous holes. Or they may have collapsed into microscopic fragments, called *spicules*, so small as to seem to the unaided

Sponge Spicule sight like a featureless brown powder.

Though found in the older Palaeozoic rock-beds, sponges do not become abundant before the Mountain Limestone; they are also abundant in the Oolite. They appear in greatest profusion in the Cretaceous beds—the Lower Greensand and the Chalk. Whenever fossils have an everyday name, it shows that they must be found in such large numbers as to attract the attention of workmen or countryfolk. The quarrymen of Berkshire and Wiltshire call the fossil sponges 'petrified salt-cellars'.

'Rotten' Flints The quarrymen contemptuously call a flint 'rotten' when, instead of breaking with the usual conchoidal fracture, it falls to pieces at the blow of the hammer and reveals a hollow interior. Its holes may branch like the limbs of a tree and have a rough lining, this lining in fact being a fossil sponge. Or it may contain a brown powder, sometimes called 'flint-meal', which the microscope shows to consist of loose sponge-spicules. Some of these 'rotten' flints can be detected, before they are broken, by their unusual shape, which is roughly similar to that of the fossil sponge inside them. The Norwich chalk yields 'rotten' flints once filled by sponges, over a metre high; these resemble a series of cups, one inside the other and all filled with chalk—having detached the 'cups' and scraped out the chalk, the countryfolk use them as flower-pots!

Fossil Corals

Modern Corals Although no coral-reefs now fringe our islands, the general appearance of coral is familiar, from museum specimens and from curios brought by travellers from overseas. It is a mass of white or pink material, perforated by multitudinous tiny holes. It looks like a mineral, and there is little to suggest that it is really of animal origin.

Yet such it is. *Coral* is formed by myriads of tiny animals, akin to the sea-anemones of our beaches and rock-pools, but differing from them in that they produce skeletons, either inside or outside their bodies, formed of calcite. Small as these creatures are, their skeletons accumulate; they form many islands and fringe the east coast

of Australia as the Great Barrier Reef. The
'solitary' corals which do not form reefs, numerous as they are, make less appeal to the
imagination.

Coral Limestones Both types, the individual
and the reef-building, have done much to form
our rocks. As might be expected, they are found
only occasionally in sands and clays; they occur
chiefly in the limestone and indeed constitute
many limestone beds. Wenlock Edge, in Salop, is
a coral-reef 50 kilometres long.

Fossil coral

The Lower Palaeozoic rocks of Salop and North
Wales include many coral-beds, as do the Devonian Beds; the Mountain Limestone, naturally
enough, is especially rich in them. In the Mesozoic
only the Oolite contains coral reefs, though solitary
corals are scattered through the other systems.
Few corals of either type appear either in the Chalk
or in the Tertiary rocks.

Types of Corals The corals themselves differ
greatly in appearance. Some form tapering curves
like horns (those which are more twisted and irregular than the others are called by the quarry-

men 'petrified rams' horns'), but may be recognized by the radial limestone plates, called *septa,* which extend from the inner surface towards the centre (although there is no need to burden the reader with the names of the different types, one is curious enough to deserve mention—*Omphyma*). Some grow from a central stem and subdivide like the branches of a tree. Some are so tightly packed together that instead of being circular they form a pattern of six-sided prisms, somewhat resembling a stone honeycomb. Some, the 'chain corals', curve through a rock-face in a long line of loops. Some form rounded masses perforated with tiny holes.

Coral Reefs   The coral limestones throw a light on the former geography of our islands. Coral reefs grow only in warm waters, which are seldom deeper than six or seven metres. How, under these circumstances, they form great thicknesses of rock is still in dispute. Darwin suggested that this occurred during a sinking of the sea-floor so gradual that the corals were able to keep pace with it; continually adding to their reef, they keep its top almost level with the surface of the deepening water. Whatever the explanation, however, these fossil coral reefs show that the shores of our islands were once washed by shallow tropical seas.

Graptolites

By no means all fossils resemble modern animals or plants. Some are of creatures entirely extinct, and related only distantly to living types. Among these are the *graptolites*, found only in the Palaeozoic; appearing in the Cambrian, they die out in the Carboniferous. They are somewhat akin to the corals, and more so to the sea-pens, plant-like animals found plentifully around our shores; some were free-swimming creatures though others were rooted to the rocks.

Graptolites

The name graptolite (based on the Greek for 'rock-writing') is given to these fossils either because some of them rather resemble quill-pens, or more probably because others look like pencil markings. Apart from those which have been converted into iron pyrites and have a pleasing brassy appearance—some resemble coiled watch springs—it must be confessed that they are neither impressive nor interesting to casual observation. They consist of black streaks, single or in pairs, almost straight or in spiral curves; usually one side is straight or smoothly curved and the other indented like the teeth of a saw.

The graptolites are thought to be the horny skeletons of tiny creatures, each of which occupied one of these indentations. It is doubtful whether the whole graptolite formed one individual or a colony, for the occupants of the different indentations were united by a strip of living matter down the common channel.

PLATE
1

Cliffs of vertical chalk; Lulworth, Dorset. (A1324)

Chalk-cut figure—the Long Man of Wilmington.
(A2983)

Gorge in carboniferous limestone; High Tor, Matlock, Derby. (A1157)

PLATE 2

Limeburning of lower chalk; near Betchworth, Dorking. (A9927)

PLATE 3   Mouth of a cave in north-western wall of gorge; Creswell Crags. (A1180)

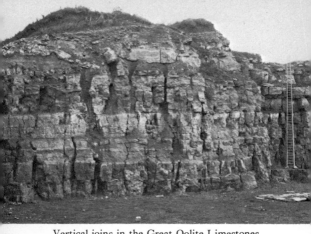

Vertical joins in the Great Oolite Limestones, widened by solutions; Kirtlington. (A3174)

PLATE 4

Thrust mass of carboniferous limestone completely surrounded by rocks of the old red sandstone; Carreg Cennen Castle, (A10779)

PLATE 5

Diorama of the Cheddar Caves. (MN2763)

Sink hole in carboniferous limestone; Bullpit, Perrfoot. (L514)

Algal limestone, nodules are the algal growths. (N1389)

PLATE 6

Unconformity of conglomeratic Upper Old Red Sandstone on Lower Old Red Sandstone (D1030)

PLATE 7

Current-bedded Bagshot Sands; Dorset. (A1418)

Basaltic pillow lava flow (lower carboniferous); Weston-super-Mare. (A11792)

Erosion of millstone grit: harder base in coarse grit capping softer bed which is being eroded more rapidly to form stacks; Doubler Stones, Rombalds Moor. (L580)

PLATE 8

Small thrust faults dipping to SE in silurian greywacke and shale (D1388)

PLATE 9

View across the Weald. (A11619)

Diorama of a Cornish China Clay quarry. (MN565)

Sharply-bedded,
steeply dipping
turbidic
sandstones, some
flaggy together
with siltstones
with dark stripy
cleaved shales.
(A11471)

PLATE
10

PLATE 11

Beds of Kentish Rag (worked as limestone). Narrow bands of hard chert alternate with soft brown calcareous sand known as 'hassock'; Quarry, Offham. (A10277)

Cliffs in blue lias strata; Kilve. (A11686)

Sharp anticline in the coal measures showing the steeper drop in the right (northern limb). An equally sharp syncline underlies this to the right; Saundersfoot, Harlow. (A10950)

Anticline in carboniferous limestone; Chepstow. (A10044)

PLATE
13

The Needles from
the East. (A1843)

PLATE
14

Monoclinal fold traversed by small thrust of lower
inclination. Carboniferous sandstone and shale
measures; Dears Door, Broadhaven. (A909)

A fold in old red sandstone; Cobbler's hole, St Ann's
head, Pembrokeshire (A11873)

Clogan Slab Quarries; Oernant. (A3125)

PLATE
15

Banded and striped hornblende and granulitic
gneiss and mica schists (Burra Voe, Shetland).
(D227)

PLATE 16

Folds in Locheil psammite (moinain); near Lough Quoich. (D1589)

Fold in lower red sandstone. (A878)

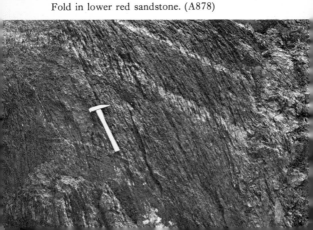

PLATE
17

Contracted mica-
schists and quartz
mica-schists;
Stunaby Head.
(NI172)

A Dartmoor Granite Tor; Blackingstone Rock. (A10059)

PLATE 18

PLATE 19

Pseudo-columnar whin overlying sandstone; Castlepoint, Dunstanburgh Castle. (A3076)

Northern parts of East Mill Tor show horizontal joints typical of the granite; Dartmoor. (A10430)

PLATE
20

Columnar jointing. (NI304)

Curved columns of Tholeiitic Lava. (NI292)

Basaltic pillow lava flow; Spring Cove. (A11792)

PLATE 21

Channel and potholes formed in calcareous volcanic rocks of the Tintagel Group by a freshwater stream; Foreshore at Trebarwith Strand. (A10326)

PLATE 22

Dyke; basalt cutting through schists; Ardna Cross Bay, near Campbeltown.

Waterfall in millstone grit Clyn-Gwyn Falls, near Ystradfellte.

PLATE
23

Mica-schists with
quartz Augen;
south-west of
Craig-Fawr.

Mountain scree at Wast Water; Lake District. (A11875)

PLATE 24

View of Stirling
showing meanders
of the River
Forth. (D1010)

PLATE
25

Head deposits overlying upper chalk; west of Cuckmere Haven. (A9953)

PLATE 26

The Northern Malverns from Herefordshire Beacons. (A11099)

PLATE 27

PLATE 28

Contorted Eelwell Limestone, Middle Limestone Group (Lower Carboniferous); on foreshore Scremerston. (A3046)

Fold in Wenlock Limestone Shales; Wren's Hill, Dudley. (A1962)

PLATE
29

Sharp Edge and
Scales Tarn with
Corrie. (G. V.
Berry)

PLATE 30

Unconformity of Upper Old Red Sandstone conglomerate and sandstones on vertical silurian greywackes and shales. (D1383)

Stromness flags (middle old red sandstone); north from Brough of Biggin. (D1550)

PLATE 31   Detail of a fossil tree—Lepidodendron trunk 26 ft+, standing nearly upright; the Cromford Canal Opencast. (L923)

PLATE
32

Close view of a colony of Litho-strotion
pauciradiale, lying close to the base of Malmerby
Scar Limestone.

Lower Old Red Sandstone conglomerate. (D1567)

Though so unimpressive, graptolites are of immense technical interest, and their nature has given rise to endless controversy. By enabling certain beds of ancient rock to be unmistakably identified, they form a valuable guide to the geological history of a district.

Fossil Worms

The humble earthworm is a geological agent of inestimable importance, helping to convert the ground in which it lives into fertile soil. So soft-bodied a creature seems unlikely to leave fossils, and most fossil worms are indeed of the marine type.

Fossil Worm-tracks

Soft though their bodies are, they leave tracks as they move over the surface of the sand; and some of these tracks persist as indelible grooves now that the beaches over which they travelled have hardened to form solid rock. Just as the ship-worm burrows into timber, so some types of worm burrow into shellfish, or even into the rock itself. The holes they make are as much fossils as the shell into which they penetrate, and are as permanent as the rock.

Other worms have formed tubes, made either of calcite or of sand-grains and fragments of shell, a little larger than their own bodies. The silt which seeped into these tubes has hardened inside them and converted them into fossils, curving over the surface of the rocks or attached to the exteriors of fossil shells.

Fossil Sea-mats

Many stones and shells on our beaches, and even some of the seaweeds, appear to be partly covered with a lace-like material. This is a *sea-mat,* a colony of tiny creatures which build cells of chitin or calcite. The cells are united in networks or separate fronds (some are easily mistaken for seaweeds), glassy-looking while their inhabitants are alive but white after their death.

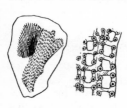

Fossil sea-mats, consisting of calcite or silica, may be found in rocks of all ages from the Silurian onwards. Like the modern type, some of them spread, in a delicate lacy crust, over fossil shells or sea-urchins. They are especially numerous in the Mountain Limestone

Fossil Sea-mats

and the Chalk, and are so abundant in the Pliocene of East Anglia that one bed of shelly sands and clays has been named after them—the Coralline Crag of the country round Aldeburgh and Orford.

Fossil Starfish

Most animals have a 'bi-lateral symmetry', the two sides of their bodies being almost 'mirror images' of each other. Some, however, have not a twofold but a *fivefold* symmetry; their bodies have five parts almost exactly alike.

Fossils with a similar fivefold symmetry are very common. The fossil *star-fish*, brittle-stars and feather-stars unmistakably resemble the modern type; though their bodies are so fragile that most of them can be found only in de-tached fragments, some are excellently preserved in their entirety. They occur most plentifully in the Leintwardine and Ludlow districts of Here-fordshire. (The collection in the Ludlow Museum is excellent.) They may also be found in the Blue Lias Beds on the coast near Lyme Regis, Dorset, and near Whitby and Staithes, North Yorkshire, on the South Wales coast, in the Yorkshire Lime-stone grits, and in the Chalk of East Anglia and Kent.

Fossil Starfish

Fossil Sea-urchins

Among the 'common objects of the sea-shore' are the round, flattish *tests* (shells) of sea-urchins.

They may bristle with spines or have lost these, the bare test resembling delicate basket-work. (The spines give the animal its name, 'urchin' being an archaic term for hedgehog.)

Though so different in appearance, the *sea-urchin* is distantly related to the starfish, and is likewise of fivefold symmetry. Its body possesses five lines of thin worm-like tubes, which it can protrude through the holes in its test until they project beyond its spines. It moves by means of   the suckers on these tubes. The five lines are so clearly marked in the fossil sea-urchins as to identify them at once, greatly though the various types differ in size and shape.

Fossil Sea-urchins

Like the starfish, the sea-urchins occur in rocks of almost all ages, in the Mountain Limestone of Ireland, North Wales, and the Peak District; in the Oolite of the Cotswolds, and in the Lower Greensand of Wiltshire and Berkshire. In the Chalk, they are so plentiful as to have been given 'folk names'; the peasantry call the more pointed sea-urchins 'shepherds' crowns' or 'shepherds' mitres', and the flatter, broader type, which are shaped so as to resemble a conventional heart, 'fairy hearts' or 'fairy loaves'. The fossil sea-urchins of the London Clay are of a different type, having lived in shallower water than those of the Chalk.

Fossil Sea-lilies

Attached to the sea-floor grow brown tubes, resembling the stems of plants. They end below in radiating thinner tubes like roots, and above in fleshy cups, like the calyx of a flower, from which spread five long arms, subdividing in the manner of petals. These are the *sea-lilies,* or *crinoids* (from the Greek for 'lily'). They are also known as *zoophytes*—but incorrectly, for this term is Greek for animal-plant, and the sea-lilies are not plants at all; in spite of their appearance they are true animals. As their fivefold symmetry suggests, they are related to the star-fish and sea-urchins; their great difference from these—and indeed from almost all other animals—is that instead of being free to move about they are permanently attached to the rocks.

Fossil Sea-lilies

Sea-lilies are among our oldest, commonest, and most beautiful fossils. Some have remained intact with their 'roots', long 'stems', and cup-like 'calyx', and are more apt than the living creatures themselves to be mistaken for plants. Most occur in limestones, as though they had flourished in water almost free from sediment; but some shales are crowded with them, as though a muddy sediment had killed and buried them almost at once.

They are found in the oldest limestones, from the Silurian onwards; in the Mountain Limestone some of them possess 'stems' of unusual thickness. They are plentiful in the Lias beds, the blue limestones and shales of Lyme Regis and Whitby; here some, converted into a pale stony material, look rather like plaster casts, others, partly transformed into iron-pyrites, have a metallic sheen.

The long stems of the fossil sea-lilies were formed of numerous joints, which readily become detached, each large enough and characteristic enough to be recognized. In some places these detached joints are so plentiful indeed that the country-people have given them names, sometimes calling them 'screw-stones' from their appearance (they are grooved not spirally, but with rings easily mistaken for spirals) and sometimes, more poetically, 'St Cuthbert's Beads'.

Fossil Barnacles

Barnacles grow in profusion on the rocks below high-water mark and on the timber supports of piers—also, as readers of sea-stories will remember, on the hulls of ships. Some are attached by 'stalks'; others, known from their shape as 'acorn shells', are 'sessile', having their bases fixed direct to the rocks. Both types have external shells, the edges of which are very sharp, but they differ completely from the other types of shell-fish. They are not *Molluscs*, like the cockle and oyster, but *Crustacea*, distantly related to the lobster and crab.

The life story of the barnacle is extraordinary.

In its immature stages it is a tiny, freely swimming creature, but as it matures it anchors itself to rock or timber and completely changes its internal structure and external appearance.

Fossils of both types are common, the stalked type appearing in older rocks than the sessile; they may be recognized by the acorn-like appearance of their shells, with or without the cast of their stalk. Some of the barnacles in the Norfolk Crag are attached to the flints by a cement containing iron-oxide.

Trilobites

Among the animals which have long been extinct is the *trilobite*—so called because its fossil body consists of three lobes. The larger fossil trilobites are conspicuous and unusual enough to attract attention and have long been noted as 'petrifications'—plain evidence to a pre-scientific

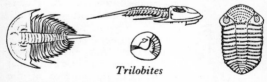

Trilobites

age of 'Noah's Flood'. In one district they are so plentiful that the quarrymen have a name for them —'Dudley Locusts'. To the casual eye they do indeed resemble some extraordinary flattened insect.

They are, however, another type of the

Crustacea, of the 'armoured', many-jointed, many-legged creatures among which are the lobster and crab. The trilobites do not greatly resemble these; they are oval-shaped, and range in size from the microscope to sixty centimetres long. Their head, usually provided with large eyes, consists of a 'shield'; the body is divided into a number of segments, and similar segments, of diminishing size, form the 'tail shield'. In some types the head shield ends in curved projections pointing down the sides of the body; in some, the body-segments bristle with legs. Grooves from the top of the head-shield to the end of the tail divide the whole fossil into the 'three lobes' after which it is named. The joints between the segments were evidently flexible, for some fossil trilobites are found rolled up, armadillo-fashion, into armoured balls.

Trilobites occur only in the Palaeozoic rocks, disappearing completely in the Permian. In these rocks, however, they are so plentiful, that almost all stages of their growth are represented, enabling the palaeontologists to reconstruct their life history.

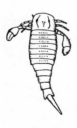

*Fossil
Sea-scorpion*

The youngest, according to the Wells-Huxley-Wells *Science of Life,* were 'tiny little tadpole things, all head and no body' that swam near the surface of the water. They grew, like many modern crustaceans, by adding segments to their bodies, and as they grew many of them changed their ways of life. Some—the earliest—lived and fed on the

sea-floor, crawling over it and extracting nourishment, somewhat like earthworms, from its organic
remains, their body armour protecting them
against attack from above. Others burrowed into
the mud, ate seaweed, or swam freely, propelled
themselves when danger threatened by a sudden
jerk of their tails. Some were spiky; some, as
already stated, could roll themselves into armoured balls; some had practicable legs and complicated eyes.

Lowly as they may seem, the trilobites were for
ages in the Lower Palaeozoic the dominant form
of life. Later, in the Old Red Sandstone of
Scotland, larger creatures of kindred type appeared, somewhat resembling *scorpions*—some
specimens are as much as three metres long. These
animals also possessed jointed bodies, though
these were so narrow in proportion to their height
as to lead the Scottish quarrymen to speak of their
fossils, either jocularly, or with an uneasy pseudo-
reverence, as 'Seraphim'.

Fossil Shellfish

LAMP-SHELLS

The palaeontologist, like the student of present-
day natural history, bases his classification, not on
external appearance, or even on mode of life, but
on internal structure. He accordingly treats two
creatures which most of us would lump together
as 'shellfish', as being altogether different.

One type consists of two unequal 'valves' (shells), each being symmetrical about its centre line; in the larger valve of most of them is a tiny hole, through which passed a fleshy stalk. These are not grouped technically with the other shell-fish (molluscs); they are called *brachiopods* ('arm-feet'), or more familiarly *lamp-shells,* from the resemblance which some of them bear to the lamp of classical times.

Lamp-shells

The lamp-shells are very plentiful in the Palaeozoic rocks, but appear more sparsely in the later beds, as though they were being supplanted by the molluscs. One type, however, the tiny shell known as *Lingula,* has lived almost unchanged through the ages since the earliest Cambrian rock-beds were formed. Though perhaps not the oldest form of life still in existence, it is the oldest of which we have definite record.

BIVALVES

The *bivalve* molluscs (shellfish) are familiar to us as minor articles of food; among them are the cockle, mussel and oyster. Though some of them are similar in appearance to some of the lamp-shells, they are quite different in structure. Their two shells, which are hinged together, may be elaborately notched, like those of the cockle, the notches fitting into one another to make their protection more complete; or they may have a smooth outline. Some bivalves like the mussel, 'anchor' themselves to the rock by threads of adhesive material produced from their body; others, like the cockle, can move about by a long 'foot' which they protrude from the shell, retracting this when danger threatens.

Bivalves

Fossils of the cockle, mussel and oyster, with those of similar creatures less familiar to ourselves, are found in rocks of almost all ages. Many can at once be recognized by their appearance. One of the unfamiliar types, known as *Pecten,* may be sixty centimetres across! Another type is common enough for the countryfolk to give it an extraordinary name, calling what are technically known as *Gryphaea,* 'devil's toe nails', in playful reference to their shape.

UNIVALVES

The lamp-shells greatly resemble the bivalve molluscs, but only in external aspect. The single-shelled molluscs, though totally different from the bivalves in appearance, are classed along with them. Many of these *univalves,* like the snail on land and the whelk in the sea, have spiral shells; the limpet's shell forms a plain shield, which its inmate can anchor almost immovably to the rocks.

The univalve fossils appear in the ancient rocks, but exist in greatest profusion in the Tertiaries. Here they are so numerous that they enable these beds to be classified according to their age by the proportion of modern types among them. (See page 93.)

The loose, crumbling sands called *Crag,* of East Anglia, are very rich in the fossils of univalve molluscs, almost all of them spiral, but varying greatly in shape and size. These specimens are very easy to collect, but have the disadvantage that to the inexperienced eye they are hard to distinguish from the shells of the modern sea-shores; they lack the distinctive archaic appearance one naturally expects of a fossil.

Univalve

AMMONITES

Nautili The *pearly nautilus* of the Indian Ocean is a mollusc of an unusual type. It is related to those forbidding monsters, the octopus and the cuttlefish; possessing tentacles with which it grasps its prey, it builds a beautifully coiled spiral shell, which is divided by partitions into a series of compartments. The nautilus inhabits these in turn. When it grows too big for one compartment it builds another larger one, thereby extending its spiral shell.

Ammonites   Fossils of the nautilus are plentiful. Still more abundant are those of a closely related creature which is now extinct. This is the *ammonite* (its name comes from the classical god Jupiter Ammon, whose head was embellished, as a sign of power, by ram's horns, to which the ammonite shells have a picturesque resemblance).

Ammonites, showing cross-section of interior

The ammonites appear in the Devonian beds, though nautilus-like forms occur as early as the Upper Silurian, and continue through the Palaeozoic and Mesozoic, dying out at the end of the Cretaceous. Especially beautiful fossils abound in the Blue Lias of Lyme Regis and Whitby, ranging in size from a metre or so across to no more than a few millimetres. Some of the larger ammonites appear only in outline on the limestone boulders, but most of the smaller ones in the dark blue shale have been transformed into calcite or iron pyrites—those in pyrites are especially fascinating, looking as though they had been cast in bronze.

Interior Structure   When split down the middle and polished, the larger ammonites not merely look attractive but enable the interior structure of their shells to be seen. Filled now with calcite, the various compartments can be recognized; they are separated by curved partitions. In the fossil nautilus, these partitions are gently curved throughout. In the ammonites, though gently curved at the centre, they fold in fantastic complications where they join the exterior shells —some of these folds can be faintly distinguished in the fossils where the shell itself has perished, leaving only a calcite 'cast'.

A 'Pious' Fraud   The spiral coils of certain ammonites suggest headless snakes. This ap-

pearance so impressed the pious folk of the Middle Ages as to evoke a legend that the fossils were indeed the remains of actual serpents miraculously petrified by the Yorkshire Saint, St Hilda. The less pious folk of the period 'cashed in' on this legend by carving heads on the ends of the spirals and selling them as 'relics'!

A 'relic'

Cretaceous Ammonites   During the last period in which they appeared, the Cretaceous, many of the ammonites had fantastic shapes instead of the usual flat spiral. Some were straight; some doubled back on themselves in a 'hairpin bend';

some were crescents; some formed spirals with spaces separating the whorls, which might also be interrupted by a length of straight shell; and some assumed a tapering helical curve like a corkscrew.

Cretaceous Ammonites

Belemnites

The octopus and cuttle-fish themselves are not what one would ordinarily think of as molluscs (shellfish), but in fact they are. True, the cuttlefish have no visible external shells—but neither has the slug, itself a mollusc, and plainly resembling that obvious mollusc, the snail, except in having its shell *inside* its body. The octopus and cuttlefish are distantly related to the nautilus and the ammonite, but their shells are similarly internal. The cuttlefish 'bone' is a fairly familiar object (specimens may be picked up on the beach, and are often placed in budgerigar cages for the bird to peck at); it consists of a thin slab of white material with curved bevelled edges, and is really an internal shell.

The analogous 'bones' of extinct types of cuttlefish occur as fossils, but only in the Mesozoic. Like the ammonites they are unknown in the Tertiaries and, like them, they abound on the Blue Lias shores of Lyme Regis and Whitby. They differ greatly from the ammonites, however. They resemble slate-pencils of unusual thickness; circular in section, and tapering at one end, they have been called *Belem-* 'cigar-shaped'. They are so common as to *nite* suggest a folk legend that they are 'thunder-bolts; indeed, their name, *belemnites*, comes from the Greek for 'dart'. These are, to be exact, only the 'guard' to the true shell, though they may be found, but infrequently, with the remains of a conical shell embedded in the blunt end.

Fossil Insects

Even creatures so frail as the insects can become fossils; flies entrapped in amber, a hardened gum, are preserved in their entirety. Those fossilized under more usual circumstances, in silt which hardened into rock, are mostly incomplete. There are few specimens with fragile wings, like the moths and butterflies, the commonest being of a 'compact' type resembling. the cockroaches and beetles.

Though a few are found in the older Palaeozoic beds, *fossil insects* first became abundant in the

Carboniferous. In this period there were giant dragonflies with a wing-span of sixty centimetres, as well as smaller types, crickets, cockroaches, fireflies, and—as might be expected in an age of forests—plant lice. In this period, too, lived other lowly creatures—spiders, millipedes and galley-worms.

Many modern insects would have been unable to live during the period, however, for the forests which gave us our coal contained no flowering plants. These do not appear until the Mesozoic, and in this era the insect-life bears a greater resemblance to that of the present day.

Fossil Fish

As water-dwelling creatures with hard skeletons, fish are well adapted to preservation. Fossil fish —the earliest fossil *vertebrates* (creatures with backbones)—appear in the Silurian, abound in the Old Red Sandstone of Scotland, and continue until the present day.

The earliest fish, though they somewhat resemble the sturgeon and the American garpike, were, however, very different from most of the modern types. They are known as *ganoids* (from the Greek for 'brightness') on account of the scales and plates of bone, shining as though they had been enamelled, with which they were protected. Some had numerous overlapping scales; in others the heads and part of the bodies were shielded by a thick bony armour. Most of these early fish lived in the lakes and rivers; their remains are comparatively infrequent among the Devonian marine fossils of the same period.

Fossil Fish

In the Cretaceous appeared such modern fish, with the familiar bony skeleton, as the cod, herring and salmon, and here, too, sharks' teeth are numerous. Some of these remains are found not as scattered specimens but as *fish-beds*, suggesting that a whole shoal of fish had been overwhelmed by some catastrophe—a sudden change of climate, perhaps, a spate of fresh water into the sea, or the liberation of poisonous gases by a volcanic eruption. One such fish-bed occurs in the Osborne Series of Oligocene rocks in the Isle of Wight.

Fossil Amphibians

Fossil *amphibians*—creatures which after dwelling in water for the early part of their lives undergo a remarkable change in bodily structure and emerge on dry land—form the only remains of backboned land animals in the Carboniferous. They somewhat resemble salamanders, but the folds of material in their teeth were so complicated as to earn them the name of *Labyrinthodonts* ('maze-toothed'). Most were small, but some

Head of a fossil amphibian

were over two metres long. The large amphibia also lived in the Triassic, and have left its sandstones marked with their footprints, but later in the Mesozoic they give place to smaller amphibia of a more modern type.

The remains of one Miocene salamander achieved a curious notoriety; they were exhibited in 18th-century Germany, and described, illustrated, or alluded to in contemporary books and sermons, as the bones of one of the 'giants' mentioned in Genesis as having lived 'before the flood!'

Fossil Reptiles

The Age of Reptiles Mention of the fossil reptiles brings us on familiar ground. Most people, if asked about 'prehistoric animals', would at once think of some monster after the style of the much-publicized Brontosaurus. Many of these creatures were, indeed, incredibly large and incredibly strange in appearance; their remains form the most impressive and most popular exhibits in our museums. They first appeared in the Permian, the latest of the Palaeozoic periods, but attained their greatest development in the Mesozoic, *the* Age of Reptiles.

Dinosaurs Some reptiles were comparatively small and commonplace, resembling the modern turtles and alligators; others were of types which have long since perished. Of these, the greatest were the Dinosaurs (from the Greek for 'terrible lizard') remarkable for their huge body, four stumpy legs, long neck with its small head containing a tiny brain, and long tapering tail. Among them are not only the *Brontosaurus* ('thunder lizard'—most of these names end in '-saurus', the Greek for 'lizard'), but the *Atlantosaurus,* the *Gigantosaurus* and the *Diplodocus*—some of these creatures were as much as seven metres high at the shoulder and little short of thirty-three metres long!

Other Dinosaurs, though smaller, were elaborately armoured with bones and spikes. The *Stegosaurus* (about eight metres long) had a double row of great vertical plates along its back, and an array of sharp spikes on its tail. The *Triceratops* had a great bony plating covering its skull and extending like a frill over the back of its neck; on the front of this plating were the three horns which give the creature its name.

Dinosaur

Brontosaurus

Stegosaurus

Carnivorous Dinosaurs These great reptiles lived on land (Plate 23)—though the larger ones may have waded partly submerged in the mud and water and browsed on the vegetation along the shores—and were herbivorous (vegetarians). There were, however, carnivorous dinosaurs, comparatively small indeed, but armed with appalling teeth and claws and presumably very fierce: the *Tyrannosaurus* and the *Allosaurus*. These had massive hindlegs but smaller forelegs, and seem to have been able to rear up, and perhaps even to run, kangaroo-fashion.

Marine Reptiles There were also giant reptiles of marine type. The *Ichthyosaurus* ('fish lizard') is so called because of its tapering body, adapted, like that of a whale or a seal, for life in the water. It possessed not fins but paddles, formed of many jointed bones, a fish-like tail, large jaws well armed with teeth and great circles of bone round its eyes. The *Plesiosaurus* ('near lizard') had a long tapering tail, and a small head on a long neck, which may have been not flexible like a swan's but have stretched rigidly in front of the body.

Flying Reptiles Much smaller, though even more extraordinary, were the *Pterodactyls* ('finger winged'). They were flying reptiles, with web-like 'wings'; these were, however, very different from those of the bat. Instead of extending between the outstretched 'fingers', the wings formed one continuous surface, joining the 'little finger' to the body.

Although the complete skeletons of such

Pterodactyl

Plesiosaurus

Ichthyosaurus (above) *and skeleton* (below)

animals have been discovered, they are unlikely to come within range of the casual collector. He may, however, find, in the Blue Lias of Lyme Regis, for example, an isolated bone or segment (vertebra) of the spine.

End of the Reptile Age None of these monsters are found in the rocks later than the Chalk. They seem—though the smaller and more 'commonplace' reptiles survived—to have vanished completely from the earth. Some great change in climatic conditions may have destroyed them, together with the ammonites in the sea. Or the flesh-eating reptiles may have killed off the herbivores, and then perished of starvation. Or much smaller but more intelligent creatures, the forerunners of the mammals, may have exterminated even the largest or fiercest reptiles by eating their unhatched eggs.

Fossil Birds

Though birds seldom became fossils, some extraordinary specimens have been preserved. They show a distinct resemblance to reptiles! (Though perhaps this is not so very extraordinary considering the scaly covering of a hen's legs.) They are, however, quite distinct from those flying reptiles, the Pterodactyls.

The earliest, found in a fine-grained Jurassic limestone known from its use in printing as 'lithographic stone', is the *Archaeopteryx* ('ancient winged'). It differed from modern birds in having teeth, a long tail, clawed fore-limbs and other reptilian features; it might indeed have passed as a true reptile had it not been feathered, not only on its three-clawed wings but down the sides of its reptilian tail. It is generally regarded as intermediate between reptile and bird—an example of what was once called a 'missing link'.

Archaeopteryx

Two later specimens also possessed teeth, though in other respects they were unmistakable birds. *Ichthyornis* was a water-bird able to fly; *Hesperornis,* also a water-bird, had only rudimentary wings and was a 'diver'. These lived in the Cretaceous, in which also appeared birds of a more modern type, related to the ostrich. The observer is extremely unlikely to find such interesting specimens in the rocks, but he may study them in the geological museums.

Fossil Beasts

Beast-like Reptiles Much as Archaeopteryx
was intermediate between reptile and bird, so
other Mesozoic reptiles showed some features
characteristic of the *mammals* (beasts). Among
these was the *Theriomorph* ('In the form of a
beast'). Though other early beasts appeared in
the Age of Reptiles, most of them are small and
insignificant; their fossils are to be studied not in
rocks but in museums.

The Age of Mammals The Caenozoic was,
however, the Age of Mammals, in which beasts
were the dominant form of life. Some of the
earlier Caenozoic beasts, though neither so large
nor so fantastic as the giant reptiles, were monstrous in size and form. The *Titanothere* and the
Unitathere were clumsier than the modern hippo-
potamus and had even more ugly heads and horns.
The *Entalodont* was a giant pig, almost as high at
the shoulder as a man. Later these were replaced
by creatures less monstrous but no less strange:
the *Oxydactylus,* for example, combined the
features of the modern giraffe and camel.

Eohippus One interesting sequence of creatures began with the *Eohippus,* an animal hardly
bigger than a fox, with the five-toed feet so
characteristic of most beasts and with teeth of a
simple type. Its successors grew progressively
larger, their teeth became more complicated, and
they shed their toes one by one. Their modern
representative runs with great swiftness on what

134

was once the middle toe and is now a single-toed hoof; it is, in fact, the horse.

Early Elephants   Another sequence includes a number of beasts of unusual massiveness: the *Tetrabelodon*, the *Dinothere*, the *Mastodon* and the *Mammoth*. Their faces and jaws became progressively shorter ('bull-dogging', this has well been called); their upper lip extended to become a prehensile trunk; two of their teeth developed into tusks. Their modern representative is the elephant.

Mammoth

The Great Ice Age   Many of the Pleistocene beasts show clearly that they lived during that time of intense cold, the Great Ice Age. They include creatures long extinct, like the sabre-tooth tiger, the woolly mammoth and the woolly rhinoceros; others, like the reindeer and the bear, no longer native to our milder climate, though they still flourish in the bleak north regions; others formerly very numerous though now they have almost died out, like the bison which lives in the great forests of eastern Europe; and some more ordinary creatures, like the little shaggy pony, the wild goat, and the 'izard', the Pyrenean chamois.

Fossil Man and His Forerunners

The Sub-man It was during the Great Ice Age that human beings first appeared. They were, however, preceded by creatures in some respects human, in others more akin to the ape. The actual remains of such *sub-men* are very scarce, but specimens of the flint implements which they made are abundant.

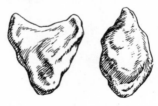

Eoliths

Eoliths The earliest of these are so crudely shaped that a controversy still persists whether they were indeed the products of human handi-work, or whether they were produced solely by natural forces. These *eoliths* ('Dawn stones') were first systematically collected and studied by

Benjamin Harrison, of Ightham in Kent. He found them plentifully scattered on the North Downs near Maidstone. They consist of flints very roughly chipped indeed; the creature which made them is thought to have been a very primitive type of sub-man.

Neanderthal Man More abundant remains have been discovered, though not in Britain, of another sub-human creature known as *Mousterian* or *Neanderthal Man*. Its head was shaped

Late Mousterian implements—borer, lance head, and scraper

differently from the human, with receding fore-head and jaws, and with great brow-ridges over the eyes; its teeth and thumbs were also mis-shapen; its back was bent so that it could not walk upright, and it may have been covered with a coat of bristly fur. It made cleverly chipped 'fist flints', pointed at one end and shaped to grasp at the other. It may be remembered in our folklore, for such creatures, grotesquely semi-human, dwelling in caves and lairs, and preying on human flesh, may have suggested the gnomes and trolls and ogres of our fairy tales.

Early Man The actual bones of the earliest human beings are far more scarce than the stone implements which they made. These include spear-heads, axe-heads, knives, scrapers, borers, and saws, beautifully chipped from flint; also bone

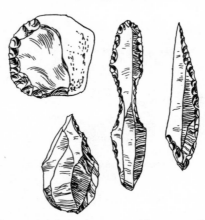

Early man-made implements — side-scraper, burin, scraper, knife

and ivory harpoons, fish-hooks and ornamental carvings. The most fascinating relics of these early men are the paintings they made on the walls and ceilings of their cave dwellings in France and northern Spain, depicting such

Drawings from the French caves

creatures as the mammoth, woolly rhinoceros, bison, goat and pony.

But enthralling as it is, the study of primitive man is a matter less for the geologist than for the 'pre-historian', and so is beyond the scope of the present book.

*Creatures of the Great Ice Age, as drawn by pre-
historic man. The Horse's Head is from Cresswell,
Derbyshire; the others are from the French caves*

MASSIVE ROCKS

Granite

Character In the more rugged parts of Britain are masses of rock completely different from the bedded types already described. They are, with few exceptions, not arranged in parallel layers; they are not composed of pebbles, sand grains or hardened mud; they contain no trace of any fossil. They present not the slightest indication of having been formed, as the bedded rocks were formed, through the slow hardening of deposits on the floor or beaches of sea or lake.

Occurrence The commonest of these massive rocks is *granite*, familiar almost everywhere if not as a building material at least as a stone favoured by the 'monumental mason'. It occurs in Dartmoor, in the cliffs of Land's End and elsewhere in Cornwall; in the Lake District; in Lleyn (western Gwynedd); in the Aberdeen region and in other parts of Scotland.

Structure To a casual glance, granite appears to be speckled white or grey or pink; it is heavy and rough and hard to the touch. Closer inspection reveals it as a jumbled mass of tiny crystals, some almost too small and some large enough to be visible; it may also include larger crystals several centimetres long; and at least three entirely different types of crystal are distinguishable.

The predominating type, which gives the rock its colour, consists of *feldspar* (from two German words meaning 'field crystal'). This mineral may be white or pink, and its crystals are mostly small, but some may be quite large.

The 'sparkle' in the granite comes from specks of *mica* (from the Latin 'shining'). Although so tiny, and black instead of transparent, they are similar to the substance used to form the 'windows' of oil-stoves and the eye-pieces of goggles.

Intermingled with the feldspar and the mica are crystals of a grey substance, *quartz* (a term derived from the German, but of doubtful origin). Different as it appears from the flint in the chalk and the grains of sand on the sea-shore, this is a form of silica. Its crystals are more irregular than those of the feldspar and mica, and look as if it had filled the spaces between them after they had been formed. There may also be smaller quantities of other minerals, ranging from tiny dark-red garnets to magnetic iron ore.

The granite shows no other signs of internal structure, the formless tangle of crystals persisting throughout its mass. Here and there, however, it may be broken by a cavity; into this project crystals larger and more regularly shaped than those which form the rock itself. Granite in which such cavities are numerous is called *drusy*. Here and there, too, it may be traversed by veins of ore or other minerals; these show nothing resembling bedding but may slope in any direction.

Uses The hardness of granite makes it suitable for such purposes as road-making. Combined with its attractive appearance—the rock can be smoothed and polished—its hardness makes it no less suited for the construction of dignified and durable edifices and monuments. It was used for such purposes in ancient Egypt, as it is today, and the buildings then erected still impress us with their majesty and strength.

Although granite is not a bedded rock, its surfaces, projecting above the ground in the *tors* of Cornwall, may give a misleading impression of bedding. It is traversed by perpendicular cracks, caused by its contraction as it cooled, and the weather, acting on its upright cliffs, wears the cracks into grooves, producing a delusive resemblance to worked courses of masonry; this is called *mural* ('wall-like') jointing.

Formation of Granite In everyday experience, crystals are formed when a solution of sugar or alum, for example, is boiled away or evaporates; and the more slowly it does so, the larger and more perfect are the crystals. They can also be formed, however, when a molten material cools down and solidifies—provided that it does so slowly enough.

Granite is an example of an *igneous* (Latin 'fiery') rock, one formed by the cooling of a mass of molten material. The high temperature at which it melts shows the conditions of intense heat under which it was formed, and the size and perfection of its crystals show the extreme slowness with which it cooled.

A Granite Boss   The arrangement and appearance of the adjacent rocks indicates that volumes of molten material welled up from the intensely heated interior of the earth. They did not reach the surface, however; instead, they pushed up the upper layers of the bedded rock to form 'blisters', and welled into the dome-shaped spaces thus formed. There, protected by the layers above from the chill of the air, the molten rock was able to cool so slowly as to produce visible crystals. Thus it formed a dome-shaped mass of rock which became exposed at the surface only when the rocks above it were destroyed by the weather; this is called a granite *boss*.

Formation of a Granite Boss

Action of Weather   Once exposed, the granite itself was attacked by the weather, with surprising results. These do not end with the widening of the joints in the rock and the dislodging of fragments from its surface. Acted on by the chemicals in the air, the feldspar itself slowly rots and softens. Unaffected themselves, the crystals of quartz and mica loosen and fall away, are washed by the rain into streams and carried down in the rivers to the sea; rounded by friction with one another and discoloured, they become grains of sand which, accumulated on the sea-floor or cast upon the beaches, may be welded into a bed of rock. (It is in this sense that, irrespective of their geological age, the massive rocks are said to be 'Primary' and the bedded rocks 'Secondary': the one is made up of fragments or particles detached from the other.) As for the feldspar itself, from being a constituent of one of the hardest rocks, granite, it ends up as one of the softest of materials—china clay.

Granite Scenery So hard a rock as granite naturally produces bold, rugged scenery. Dartmoor covers the weather-worn surface of a huge granite boss, and Bodmin Moor that of another (Exmoor lies not on granite but on the Devonian bedded rocks). These 'forests' are unsuited for cultivation; they consist of stretches of wild country, clothed in furze and heather, broken by peat-bogs and by hills crowned with the wall-like granite tors. They can be dangerous in winter because of the morasses which feed the streams, and in summer by reason of the vipers which shelter in the heather or bask on the sun-baked rocks.

Land's End forms the remains of another granite boss not merely weather-worn but partly destroyed by the sea. Its towering cliffs seem to defy the onslaught of the fiercest Atlantic breakers; and though, hard as they are, they are none the less being slowly destroyed by the encroaching waves, yet the destruction, unlike that of the softer rocks to the north and east, is imperceptibly slow.

Other Coarse-grained Rocks

Porphyry The granites of different regions vary not only in colour but in composition and texture, and may contain more than traces of minerals other than feldspar, mica and quartz. They 'shade off' imperceptibly into other rocks so different as to demand special names. A granite —or indeed any other rock—in which a number of large crystals is conspicuous is called *porphyry* (after a Greek word for 'purple' originally applied to an ornamental Egyptian rock), and is said to show *porphyritic* structure.

Syenite Among the rocks which somewhat resemble granite is one which contains little or no quartz, and is formed chiefly of feldspar; it may also contain mica and other dark brown or black minerals, with perhaps a suggestion of green, known as *augite* and *hornblende*. This rock, which may be either grey or red but which is rather darker than granite, is called *syenite* (after Syene in Egypt) and occurs in Gwynedd and in Scotland.

Diorite Another dark rock also resembles granite except that its grains are smaller; it contains feldspar, hornblende and mica, and may also contain a little quartz. This rock, *diorite,* may be dark grey or show the mottlings of green and white which have led to its being known also as *greenstone*. It occurs in the Highlands and in the south of Scotland.

Gabbro The dark rock called *gabbro* owes its colour to the minerals feldspar and augite; it also contains a pale green mineral appropriately known as *olivine*. It is found near the Lizard; near St David's, Dyfed; in the Lake District, Cumbria; in the Isle of Skye and elsewhere in Scotland.

Basalt

Occurrence Of all the geological formations in these islands, surely the most striking is the columnar structure of Fingal's Cave in the Island of Staffa and the Giant's Causeway on the coast of Northern Ireland. The rock in these regions is *basalt* (derived through Latin from an African word). In appearance, as in structure, it differs completely from granite. This rock is also found in thin sheets interstratified with the limestones of Matlock and of other parts of Derbyshire, and occurs in the North Country and elsewhere.

In composition, basalt greatly resembles gabbro, consisting of feldspar, augite, and olivine; its great difference lies in the smallness of its crystals. Those which form the granite and the other coarse-grained rocks are large enough to be plainly visible. To the unassisted eye, basalt appears as a black compact rock; except for large porphyritic crystals here and there it does not appear crystalline at all. But when a slice of the rock, so thin as to be translucent, is examined under a microscope, it reveals itself as a mass of tiny crystals welded tightly together.

The Giant's Causeway

Formation This contrast in size of the crystals in the two rocks is due to the different circumstances in which they solidified. The coarse-grained rocks were formed well underground; though also the result of an uprush of molten material from below, the basalt actually reached the earth's surface. Chilled by contact with the air and the ground, it hardened very quickly, with the result that its crystals are microscopically small.

Though some specimens look very handsome when worked and polished, unworked basalt is not so attractive as granite. In shapeless masses, it seems featureless; it is the sort of thing one regards vaguely as just 'stone'. It has no internal structure, and from its nature it cannot possibly include any fossils.

Columnar Structure When, however, as in Staffa and the Giant's Causeway, it forms a series of prismatic columns, it is both interesting and picturesque. They are due, like the cracks in the granite, to the rock having contracted as it cooled. They form a fascinating 'three-dimensional mosaic', fitting compactly together. The columns have from three to eight sides (six-sided columns are usual) and are divided by horizontal joints into sections over a metre long. The shorter columns over which the visitor walks suggest a tessellated pavement; the longer ones when seen from a distance have a fantastic resemblance to the pipes of a natural organ. The scenery they produce is extremely impressive, especially when, as in Staffa, they are penetrated by caverns; and the booming of the waves in these caverns produces a weird sound which might romantically be compared to organ music. Fingal's Cave inspired the composer Mendelssohn to write one of his most famous works.

Other Fine-grained Rocks

Rhyolite, Trachyte, Andesite Basalt may be regarded as a fine-grained gabbro, and the other coarse-grained rocks similarly have their equivalents. A rather uncommon rock which resembles granite, called *rhyolite*, is also white or pink, but a microscope is needed to disclose its crystalline structure. The equivalent of syenite is a light-coloured rock, *trachyte*; that of diorite is *andesite*, which may be light or dark according to its composition. These rocks may, however, include a few crystals large enough to justify their being called porphyritic.

These rocks usually reveal their structure only under a microscope. In some places, however, they show a tendency to form prisms, rather like the basalt columns of Staffa but far smaller.

Obsidian In all these fine-grained rocks the structure is crystalline. One igneous rock, however, cooled under such conditions that it has no trace whatsoever of any crystalline structure. It resembles dark bottle-glass, and is so hard that it has been worked, like flint, to make sharp-edged tools and weapons. This rock is known as *obsidian*.

Extinct Volcanoes

Active Volcanoes Even though we may not have seen one, we all know what a volcano is. Its crater, a hole in the earth's crust, steams and smokes and emits noxious vapours, clouds of ash and streams of glowing molten rock, called lava. The ash and the lava accumulate to form round the crater a hill of typical conical shape.

Evidence of Volcanic Action Our islands contain no active volcanoes, but in many regions the rocks include material unmistakably volcanic in origin. Layers of volcanic ash have been cemented into solid rocks. To show their similarity to, and difference from, ordinary water-formed conglomerates, they are called *volcanic agglomerates*. Rounded nodules embedded in the rocks are the cooled remains of blobs of molten material and glowing stone hurled from the seething crater; these are known as *volcanic bombs*.

Lava Interbedded among the stratified rock-beds are layers of hardened *lava*. It consists of a material resembling basalt or one of the other fine-grained igneous rocks. It may disclose a crystalline structure under the microscope, or be non-crystalline and glassy, like obsidian. It may have been given a porous structure by bubbles of gas in the molten lava, like the familiar substance, itself of volcanic origin, pumice-stone.

The surfaces of a lava-flow quickly part with the hot water-vapour with which the lava is inter-mingled, and become cold while the interior is still glowing hot. The difference is apparent even when the whole of the lava has cooled; the interior resembles basalt, but the two surfaces are rough, cinder-like and 'slaggy'.

Once formed, the hardened lava is as much subject as the bedded rocks to earth movements. Formed on land, it may sink below the sea, be buried by a thick mass of sediment, lifted again to become part of the land, and exposed by the weather. The lava then appears among, and parallel to, the adjacent beds of stratified rocks, but may be distinguished by its appearance: by its lack of internal structure, by its slaggy surfaces, and by the alteration which its heat produced in the rock immediately below—'baking' it and filling its depressions and crevices—though *not* in that above, which was not deposited until after the lava had cooled.

Volcanic Necks Much of the lava in the volcano does not overflow but remains in the crater. When the volcanic activity ceases, this lava cools, plugging the crater with a solid rock so hard that even a fresh volcanic outburst may be unable to dislodge it but may force its way through the earth elsewhere. This solidified lava remains after the volcanic cone is destroyed by the weather, projecting above the ground. Yet hard as it is, it none the less slowly weathers to form a steep-sided hill of characteristic shape, called a *volcanic neck*. Several such necks occur in the Scottish Lowlands: the most striking is that crowned by Dumbarton Castle in the Forth of Clyde; two dominate Edinburgh as Arthur's Seat and the Castle Hill; and another forms North Berwick Law in Borders.

Dykes and Sills

The molten material thrusting upwards from within the earth may neither reach the surface as a lava-flow nor displace the upper rock-beds and accumulate between them to form a granite boss. It may rise, partly forcing and partly burning its way, through weak places in the bedded rocks; it scorches and bakes them and percolates into their crevices, but the contact chills it, and cools it so quickly as to transform it into a thin sheet of fine-grained rock. Where the molten material rose upwards, the sheet will form a vertical *dyke*; where it forced its way between the layers it will run generally parallel with them as an intrusive *sill*.

A dyke is unmistakable, not only on account of its composition, but because it cuts right across the layers of the bedded rock, standing upright among them like a wall. The sill, though it lies among, and parallel to, the bedded rocks, is hardly likely to be confused with them; its composition is too different. It might be mistaken for a lava-flow except that the two, though generally similar, have several noteworthy differences.

Unlike a lava-flow, a sill does not have slaggy surfaces. It may 'transgress' the bedding of the adjacent rocks, rising through a weak place in a rock-bed and resuming its former direction on the other side. It 'bakes' and alters the rocks which it touches and sends 'veins' into their crevices, both above and below; the lava-flow has this effect only on the rock-bed beneath.

These layers of igneous rock, being mostly harder than the surrounding bedded rocks, produce striking effects on the scenery, forming ridges standing out in bold relief. The Roman Wall runs for part of its length along such a ridge, taking advantage of its obvious defensive advantages. This ridge is formed by the Great Whin

Sill, which stretches half across the north of England. Almost as impressive is that formed by the Cleveland Dyke, of Yorkshire. The scenery produced by such rocks is especially picturesque where the rivers cross them, tumbling over their edges into valleys worn in the softer bedded rocks, and producing *forces* (waterfalls) of great beauty.

Transformed Rocks

Metamorphism   Great masses of molten material do not force their way through the rocks without affecting them greatly. Sections of the earth's crust are not moved bodily, contorted and ground together, without producing marked effects on the rocks involved. Rocks subjected either to intense heat or to violent earth-movements are altered so completely as to demand a special name. They are called, though not very happily, for the alteration is not so much in outward form as in structure and chemical composition, *Metamorphic Rocks,* and are said to have undergone *Metamorphism* (from Greek words meaning 'change of form').

Contact Metamorphism   Round the granite bosses of Cornwall extend *Metamorphic Aureoles* ('haloes'), irregular rings in which the surrounding beds have been transformed by intense heat and pierced by tapering veins of granite. The effect is most strongly marked immediately around the boss, and becomes decreasingly perceptible farther away, so that the aureoles blend gradually into the unaltered rocks around. They are produced not only by heat, but by chemical action, jets of steam or boiling water or molten or dissolved minerals having been forced into the rocks.

Earth Movements The effects of earth movements may be seen to greatest advantage in Scotland, where they affect great thicknesses and wide areas of rock. So diverse were they, and so complicated are the structures they have produced, that the geologists have not yet accounted for them or even described them fully.

Deformed Fossils Fossils, as might be expected, are seldom found in metamorphic rocks, having been destroyed or mutilated beyond recognition by the pressure and heat, or even transformed into smears or beds of graphite. Of the few that survive, some are strangely deformed by the shearing action of the moving rocks.

Gneiss A common effect of metamorphism is to arrange the minerals in the rock, grouping them not in parallel layers but in irregular bandings. *Gneiss* (the term, pronounced 'Nice', comes from an old German word for 'sparkling') consists, as does granite, of quartz, feldspar and mica, and may indeed be a metamorphosed granite. Its constituent minerals have been, however, so to speak, 'sorted out' into bands formed of crystalline grains, alternately light and dark. It may also include tiny crystals of a dark-red semi-precious mineral, *garnet.*

Schist In a *schist* (named from the Greek for 'split'), the minerals are grouped in flaky layers, the flakes of which can easily be detached. When much mica is present, the flakes have a silvery sheen, and the rock is called a *mica-schist*. It may also contain quartz, and the crumpled nature of its layers shows what intense compression it has suffered.

Marble More familiar among metamorphic rocks is *marble,* a limestone in which the calcite has become re-crystallized. Except when dis-coloured by impurities, it is pure white, and its beauty leads to its use in sculpture and in orna-mental building work. (The term 'marble' is also loosely—and quite incorrectly—applied to some ornamental limestones.)

Serpentine Magnesian limestone, olivine, and some other rocks, may be transformed into *serpentine*. This is a softish rock, mostly green in colour, but in some places showing red mottlings; its more attractive varieties are used as an orna-mental building stone. It constitutes the bold cliffs which form the Lizard Headland, Cornwall.

Quartzite A sandstone or bed of sand exposed to intense heat may become a form of *quartzite,* remaining unaltered in composition but having its constituent grains melted and re-crystallized. It becomes so hard as to lose its structure; instead of splitting most easily along its layers, it breaks in all directions into surfaces with a gleaming 'greasy' appearance.

Slate One metamorphic rock is our best-known roofing material. Exposed to great pressure shale no longer splits along the layers in which it was deposited; instead it *cleaves* into thin sheets in a direction at right angles to that of the pressure. It has become a *slate*.

MINERALS

Occurrence Almost devoid of fossils, the massive rocks offer to the collector and observer a much more varied selection of minerals than do the bedded rocks. All rocks indeed consist of minerals, but mostly these are so mingled with impurities as to be almost unrecognizable and quite unattractive. In their pure state many minerals are both interesting and beautiful.

Crystals In favourable conditions, most minerals form *crystals,* geometric solids of perfect regularity, great beauty and much technical interest. There are a number of crystal 'forms', classified by the shape of their faces and the angles between them; some show 'twinning', two crystals appearing to grow out of one another. Their systematic study forms a distinct branch of science; their value to the outdoor observer lies in the help they give him in identifying the various minerals.

Recognition of Minerals The minerals may also be recognized by their *cleavage,* the directions in which they split, which need not necessarily be parallel to their faces; by their colour and that of their powder and of the *streak* which they leave on paper (these again need not necessarily be identical); by their weight and by their hardness. A few minerals have a characteristic magnetism, smell, 'feel' or action on light. Many are most frequently found in certain rocks.

Though each mineral has its characteristic specific gravity (weight as compared with that of water) the observer will probably be content to regard them as 'light', 'medium', 'heavy' or 'very heavy' (specific gravity below 3, about 3 or 4, 5 to 6, and over 6). Similarly, though the systematic geologist has a *scale of hardness* in ten grades from talc to diamond, the observer may class them simply as 'soft' (can be scratched with the finger-nail), 'medium' (can be scratched with a knife but not with the finger nail) and 'hard' (cannot be scratched with a knife).

Every mineral has a definite chemical composition, which is ascertained by the blow-pipe (a fascinating branch of indoor study) and by other laboratory tests. (The only chemical used 'in the field' is the acid-bottle, containing weak hydrochloric acid, for testing calcite and other carbonates.) The value of the minerals in every-day life and industry cannot be over-estimated, for they are the source of all implements and materials except those of animal or vegetable origin.

Of the hundreds of minerals in nature, there is space here to deal only with those which are common, useful in industry, or otherwise specially interesting. The first to be described are those which contain that versatile element silicon; these are followed by the rest of the type which, for want of a better name, used to be called 'earthy'; with few exceptions, they are light-coloured and light in weight. Last come those which are heavier and darker and have a metallic look; they are, in fact, the *ores* (a term probably derived from the Old English for 'brass' or unwrought metal) from which the metals are obtained.

Quartz
(Silica, Silicon dioxide, SiO_2.)

The Most Common Element Of all the elements in the rocks, the most abundant is silicon (from the Latin for 'flint'). It may constitute over a quarter of the earth's crust. It is classed chemically with carbon, the element found in all living matter, and its power of entering into complicated combinations with other elements is second only to that of carbon itself. Even as a simple oxide it is very common.

Occurrence Silica, stained with the characteristic red or yellow of iron, forms the sand and the sandstones. In its non-crystalline state it produces the nodules of flint in the chalk and the lumps of chert in the limestone. It is the material of which many fossils are constituted. Crystallized, as *quartz*, it helps to build the granite and several other massive rocks.

Quartz Crystals Quartz may be colourless and highly transparent, white or variously tinted by impurities. It may occur in shapeless lumps or form beautiful six-sided crystals with pyramidal points; some are marked with fine parallel lines, called *striations*. It is light, but may be recognized by its unusual hardness. It is in fact one of the hardest common minerals (most of the very hardest are precious or 'semi-precious' stones), far too hard to be scratched with the knife blade. In many massive rocks it is confusedly mixed with other minerals, but it also occurs separately in veins and cavities in such rocks (Plate 32), or inside *geodes* (rounded hollow stones lined internally with crystals).

Uses It is used in industry not only in the form of sand, for building, glass-making, and as an abrasive (a purpose for which its hard, sharp grains are admirably fitted), but in its other forms in pottery and as a flux.

Ornamental Stones Many of the varieties of quartz are so attractive as to have been given special names; a few are valued as ornamental or 'semi-precious' stones. *Rose quartz* is a delicate shade of pink; *milky quartz* owes its whiteness to the inclusion of very many air-filled holes. *Chalcedony* is purple-grey or brown; *amethyst* is violet; *cairngorm* is also called 'smoky quartz' from its appearance. Unlike these varieties, which are translucent, some are opaque; *jasper* is dark red; *agate* consists of concentric ovals of variously coloured chalcedony; in *onyx* similar layers are straight. The peculiar "play of colours' (impressively known as *chatoyancy*) which gives *cat's eye* its name is caused by the inter-

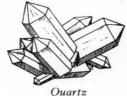

Quartz

laced fibres of which it is formed.

Rock-crystal   Perhaps, however, quartz is most attractive when it is transparent; it is then called *rock-crystal,* and is so perfectly clear that it can be used in optical work. It was known to the Greeks of classical days, who mistook it for water so intensely frozen that nothing would thaw it! This belief gave us the term *crystal,* which, though derived from the Greek word for 'ice' is now applied not only to rock-crystal itself but to any substance which naturally takes a regular geometrical form.

Feldspar

(Silicates in varying proportions of potassium, sodium, calcium and aluminium: thus orthoclase feldspar is   a potassium aluminium silicate, $K_2O.Al_2O_3 \, 6SiO_3$.)

Feldspars of different types form part of most of the igneous rocks. They constitute, for example, the white or pink material in granite. Of their many varieties, two only need be mentioned here.

Orthoclase feldspar is so named because it *cleaves* (splits) at right angles. It is pink or white, not quite as hard as quartz but about as heavy. It occurs inextricably mixed with other minerals in granite and syenite and some forms of lava, and also forms large, well-shaped crystals, oblong with pointed ends, some of which show twinning.

Plagioclase feldspar, so called because the

two directions in which it splits are not exactly at right angles, is grey or almost white. One of its varieties, *labradorite,* shows a beautiful 'play' of blue or green colour when seen at the proper angle. Plagioclase occurs in many igneous rocks, and forms prismatic crystals, some consisting of many twins side by side like thin plates, which make their faces appear to be marked with fine parallel lines. It is slightly harder and heavier than orthoclase.

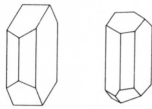

Feldspar

Feldspar is used in manufacturing porcelain and pottery, in glazing earthenware and for other industrial purposes. Under the action of the weather it decays into that very useful substance, china clay.

Mica

(Silicates in varying proportions of magnesium, aluminium, potassium and hydrogen: thus *muscovite* mica is an aluminium, potassium and hydrogen silicate, $2H_2O.K_2O.3Al_2O_3.6SiO_2$.)

Mica in its different varieties occurs in many massive rocks; for example it forms the black glistening specks in granite and the dark bands in schist. One of its commonest forms is white mica, also called *muscovite* mica because it was much worked in Russia. All the micas are light in weight and so soft that they can be scratched, though with difficulty, by the finger-nail. They form six-sided crystals.

Mica

The cleavage of mica is extraordinary; it splits very readily into very thin layers, some miners indeed referring to the largest crystals as 'books' of mica—they appear to consist of tightly packed 'leaves'. Mica is also unusually elastic. As its layers are translucent, they are used commercially where semi-transparent material is required, for example, in the 'windows' of oil-stoves; they formerly served, instead of glass, to make lanterns. As mica is a non-conductor of electricity, its thin layers are also in demand as insulators; they help to form the condensers of radios. Ground into powder, the mineral has other industrial uses as in the manufacture of rubber tyres and of certain lubricants—and to give the 'frost' effect on Christmas cards.

Other Silicates

Of the many other minerals into which silicon enters, a few are common enough to demand attention. Some have already been mentioned as constituents of the massive rocks.

Augite   *Augite* occurs in basalt, syenite, gabbro and several other such rocks. It is a silicate varying in composition of calcium, magnesium, iron and aluminium (an example is $CaMg(SiO_3)_2$ $(Mg,Fe)(Al,Fe)_2SiO_6$). It is a black or greenish-black mineral, of medium hardness and light weight, and may form eight-sided crystals ending in two faces sloping like a roof. It cleaves in two directions almost at right angles.

Hornblende   *Hornblende* occurs much as does augite, and is too complicated a silicate of calcium, sodium, aluminium, magnesium and iron to be represented in a formula. It resembles augite in weight and hardness, but its dark green colour has a brownish tinge; it may be distinguished by its crystals, which are six-sided and end in three faces forming a blunt angle, and by its two cleavages, which also meet at a blunt angle.

Hornblende

Asbestos   Because of its singular power of resisting heat, *asbestos*—a form of hornblende—is perhaps the best-known silicate. Its crystals are so long and fibrous as to suggest—wrongly—that it is some sort of fabric. Like the other varieties of hornblende, it is one of the minerals known as *amphiboles;* one of its varieties is a complicated silicate of calcium, magnesium and iron.

Olivine *Olivine,* so called from its pale-green colour, which also occurs in gabbro and other massive rocks, is comparatively simple in composition, being a silicate of iron and magnesium $(2(Mg,Fe)O.SiO_2)$. It is somewhat harder and heavier than augite and hornblende and is more common in shapeless masses than in crystals. In many rocks it is wholly or partly transformed into serpentine. In its crystalline form it constitutes the precious stone *peridot.*

Serpentine *Serpentine* is an hydrated magnesium silicate $(3MgO.2SiO_2.2H_2O)$ formed by the alteration of olivine, augite or hornblende; with veins of these minerals it forms the rock also called serpentine. Its name is due to its veined and mottled appearance, which includes varied patterns of white, green and red markings on green, black, red or brown. It has a greasy appearance and may feel slightly soapy; it does not form visible crystals. As it can be easily cut and turned on the lathe, it is in demand for ornamental work.

Garnets *Garnets* are small crystals, the silicates of such metals as calcium, magnesium, aluminium, manganese, iron and chromium. Some are 'semi-precious' stones, and the more valuable of them are deep red and transparent; the more common forms are opaque and brownish-red and are used as abrasives. They are of medium weight and hard, and occur in several massive rocks.

Garnet

Calcite

(Calcium carbonate—CaCO$_3$.)

Occurrence   Apart from the compounds of silicon, calcite is the commonest mineral. Slightly impure, it forms the chalk and, more discoloured with impurities, the limestone. It is also the material of which most shells and other fossils are formed. Pure, or slightly tinged with iron, it appears very different from either the limestone or the chalk. It is then a glassy substance, colourless and transparent or tinted light red, yellow or brown. It breaks readily into small slabs, their angles not square but slightly askew.

Calcite Crystals   Its crystals are six-sided, and are so numerous that the quarrymen have given them everyday names. 'Nail-head spar' is form-ed of short squat crystals, or of prisms (resembling six-sided pencils) ending in rhombohedrons; 'dog-tooth spar', of long taper-ing crystals; 'rhomb spar', of flat tablets; and 'paper spar', of very thin crystals.

Calcite—'Nail-head' (top) and 'Dog-tooth' (below)

Calcite greatly resem-bles quartz in colour, weight, and crystal form, and the two might easily be confused. They can soon be distinguished, however; calcite 'fizzes' at

167

the touch of acid and is far softer than quartz, being easily scratched with a knife.

Dripstone *Dripstone* is the form of calcite which accumulates on the rock-surfaces of caverns, hangs from their roofs in stalactites, and rises from their floors as stalagmites.

It is waxy-looking with a rounded irregular surface; it may consist of successive layers, which make a broken stalactite show concentric 'growth-rings' like those of a tree.

Iceland Spar *Iceland spar* is an exceptionally pure form of calcite, highly transparent, and with a strange power of splitting the beams of light which shine through it into two slightly diverging rays. Anything looked at through this substance appears double, being seen separately by each of the two rays. This mineral, in fact, *polarizes* the light, limiting its vibrations, which normally take place in all directions, to one plane. For this reason two crystals of Iceland spar, when suitably treated, combine to have an important use both in geological science and in engineering and industry.

Iceland Spar

With few exceptions, crystalline substances also polarize light—though less completely than calcite. For this reason, when ground thin enough to be transparent and placed between two *Nicol's prisms* formed of Iceland spar, they exhibit the most beautiful colours, otherwise invisible, which change surprisingly when one of the prisms is rotated. The geologist uses the *analysing microscope,* fitted with two such prisms, to reveal the composition of rocks. The engineer uses it to disclose the existence of strains in metals.

Aragonite   *Aragonite* is another less common form of calcium carbonate. It is slightly heavier and harder than calcite and crystallizes in a different system; some of its crystals are twins.

Dolomite   *Dolomite* (a double carbonate of calcium and magnesium, $CaCO_3.MgCO_3$) forms the strip of Magnesium Limestone of Permian age running south-eastwards from the coast of Durham. It is a little harder and heavier than calcite and needs warm acid to make it 'fizz'. Many of its crystals are pale-brown or yellow, and their lustre has earned them the name of 'pearl spar'. (This lustre is due to their crystal faces being slightly curved, a strange peculiarity which they share with the diamond.)

Gypsum
(Hydrated calcium sulphate, $CaSO_4.2H_2O$.)

Gypsum occurs in some clays and sands, as on the Kent coast east of Herne Bay. In the Midlands it forms beds one or two metres thick. It may be colourless, white, grey, yellow or red; it is light and so very soft that the finger-nail easily scratches it. Its crystals are tablet-shaped with bevel edges, and its twin crystals may have an 'arrow-head' form.

Gypsum   Gypsum in its crystalline form is sometimes called *selenite* because its lustre is vaguely suggestive of moonlight (it does not contain the element selenium); its fibrous variety is called *satin spar* because of its 'silky' appearance. Its compact massive form when white and translucent is known as *alabaster,* and is used for indoor statuary and ornamental building, but it is too quickly affected by the weather to be used outdoors. Gypsum is important in industry as the raw material of plaster of Paris, as a 'filler' in the making of paper, crayons and paint, and as a polishing powder in glass manufacture.

Fluor-spar
(Calcium fluoride, CaF$_2$.)

Fluor-spar

The beautiful mineral *fluor-spar* (also called *fluorite*, from the Latin for 'flowing', as it melts easily) occurs in association with tin in the Cornwall granite area, and with lead in the Derbyshire and North of England limestones. It crystallizes in perfect cubes but it cleaves not parallel to the sides of the cubes but diagonally, so as to cut off their corners. It is slightly harder and heavier than calcite.

Fluor-spar may be colourless, white, purple, amethyst, green, yellow or blue. When gently heated it becomes phosphorescent, shining with a gentle greenish light. Some of its varieties appear pale green when looked through but blue when looked at; this effect, though other substances also share it, is accordingly known as fluorescence.

The variety called Derbyshire spar or 'Blue John' is valued for ornamental work because of its attractive blue or purple colour. The colourless variety is clear enough to be used in optical work. Fluor-spar also serves as a flux in metal-smelting and in steel manufacture, and as a source of hydrofluoric acid, one of the few acids which will dissolve glass and can, therefore, be used in etching.

Rock-salt
(Sodium chloride, NaCl.)

Rock-salt, infrequently called halite, is perhaps unique in being the only mineral used as a foodstuff. We think of it merely as a condiment, but that is only because we have never been seriously in want of it. Historically, it is one of the oldest articles of commerce; too heavy a salt-tax has provoked more than one riot; and cattle will travel

*Rock-salt,
showing
'hopper face'*

far to get a 'salt-lick'. Common table-salt differs from the mineral rock-salt only in that its impurities have been removed.

Rock-salt may form in shapeless masses, or crystallize in cubes; their sides may be queerly hollowed in a series of tiny 'steps' forming 'hopper faces'. It is transparent or translucent, and has a glassy lustre. When pure it is colourless or white, but impurities tint much of it yellow or brown. It is quite soft and light, and its taste is unmistakable. It is *hygroscopic*, attracting and dissolving itself in moisture from the air (it should therefore always be stored in corked bottles). Its cleavage differs from that of fluor-spar in being parallel to the faces of the cubes.

The usual method of extracting rock-salt in the form of brine has already been described (see page 65). Apart from its impurities, which have their commercial uses, the mineral not only serves as a condiment but is important industrially, in the manufacture of such chemicals as sodium carbonate, of glass and of soap.

Barytes

(Barium sulphate BaSO₄.)

Barytes owes its name (from the Greek for 'weight') and the alternative term *heavy spar*, to its weight, which is unusually great for an 'earthy' mineral. It may be colourless, white, or tinted yellow or brown or red; its streak is white. It is fairly soft. Its crystals form flat 'tablets'; some are twinned and some have a pearly lustre or a fibrous structure. A group of its crystals side by side is called *cockscomb spar*. It occurs in the mineral veins of Derbyshire (where it is called *caulk*) and the North of England, and is mined for use in the production of white paint.

Glauconite

(A hydrated silicate of iron and potassium, perhaps also containing aluminium, magnesium and calcium.)

Glauconite has a greenish colour and is very soft and light. Unlike most other minerals described, it does not occur in bulk. It forms small grains in some of the Cretaceous sand-beds of the South of England, which are therefore called 'Greensand', and in other Eocene sands, in some of which it is abundant. It is in great demand as a water-softener, and is also used as a basis for green paint.

The Ores of Iron

Iron Pyrites Of the many natural compounds of iron the most conspicuous is *iron pyrites* (iron sulphide, FeS_2). Its yellow colour and brassy lustre have deceived so many inexperienced prospectors as to earn it the nick-name of 'Fools' Gold', and may have caused the Old English word for brass to give us the term *ore*.

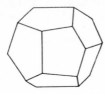

Iron Pyrites

Its crystals which may be picked up on the shore at Lyme Regis, are especially beautiful. Some are cubes, some have twelve five-sided faces, and some are 'pyritohedrons' (cubes with bevelled edges); their sides may be 'striated' with fine parallel lines. A coating of pyrites makes some fossils look as if they were coated with brass, and its yellow gleam may appear on lumps of coal. Its powder and streak are brown. Like most ores, it is both heavy and hard.

Pyrites contains too much sulphur to be worked for iron, but is used in manufacturing sulphuric acid. It was formerly used to strike sparks against flint or steel; hence its name, from the Greek for 'fire'.

Marcasite *Marcasite,* the brown nodules buried in the chalk, are another form of iron sulphide; they consist internally of crystalline 'needles', arranged radially, but quickly rust when broken and lose their silvery lustre. This mineral may also form crystals, known from their shape as 'cockscomb' or 'spear' pyrites; they are a paler yellow than pyrites itself and are somewhat lighter but are of much the same hardness.

Marcasite

Magnetite   *Magnetite* (iron oxide, Fe_3O_4) is named after Magnesia, a district in Greece where it attracted attention in classical times. It is at once the most common and the most powerful of magnetic minerals, attracting small pieces of iron and steel and playing havoc with the accuracy of the compass. A fragment of the mineral, suitably shaped and pivoted, acts as a compass; hence the alternative name of this mineral, the *lodestone*. Magnetite is black, and has a metallic lustre. It is about as heavy and as hard as pyrites. It may appear in eight-sided crystals but is more common uncrystallized. In some countries, though not in Britain, it is worked for iron.

Haematite   *Haematite* (iron oxide, Fe_2O_3) owes its name (from the Greek for 'blood') to its dark-red colour; this and the rounded masses it forms have earned it the name of *kidney iron ore*. Its internal structure is fibrous. It may also occur in black or steel-grey crystals with a lustre so brilliant that they are called *specular* (mirror-like) iron ore. Its streak is cherry-red and, like magnetite, it is about as heavy and as hard as pyrites. It is the richest ore of iron and is worked in the Forest of Dean and the North of England.

Limonite   *Limonite* (a hydrated iron oxide, $2Fe_2O_3,H_2O$) is yellow or brown in colour and streak. It is about as hard as, but rather lighter than, the other iron ores. Formed by decaying vegetation in swamps and giving their water its characteristic colour, it is called *bog iron ore*.

Ironstone　The most important ores of iron—they are rocks rather than minerals—are the clay-ironstones and the carbonates of iron ($FeCO_3$) found in the North of England. The irregular slabs of ironstone seen in the Lower Greensand hills south of London are not now worked, the richer ores of the coalfield areas being more economic. During the Middle Ages, however, they were worked intensely. Charles Kingsley, in *Hereward the Wake*, draws a vivid picture of the forges in the Weald hard at work producing the munitions which William the Conqueror used in subjugating England. They are mentioned in the *Domesday Book*, and were in use until the sixteenth century. Some of the loveliest pools here are water-logged opencast workings; some of the loveliest hills are old slag-heaps; and the name of Abinger Hammer, in Surrey, reminds us, like the Hammer Ponds, of a now forgotten forge.

The Ores of Copper

Copper Pyrites　The chief ores of copper have unusually attractive colours. *Copper pyrites,* also called *chalcopyrite* (copper and iron sulphide, $Cu_2S.Fe_2S_3$) may accompany and might easily be confused with iron pyrites. It is, however, softer and rather lighter and crumbles under heavy pressure; it does *not* strike sparks with steel or flint. It is, moreover, a deeper yellow than iron pyrites, and

Chalcopyrite

may have an iridescence which makes the miners call it *peacock ore*. It forms few crystals and is one of the most important ores of copper.

Malachite *Malachite* (hydrated copper carbonate, $CuCO_3.Cu(OH)_2$) gets its name (from 'mallow') from its beautiful green colour. It may form rounded masses like haematite, with an internal fibrous structure showing concentric bands of light and dark green.

Chessylite *Chessylite* (hydrated copper carbonate, $2CuCO_3.Cu(OH)_2$) is called after a district in France, and is no less beautiful but is blue; hence it is called *azurite*. It may form crystals like thin 'needles', radiating from a common centre. Like malachite, it is used as an ornamental stone and is worked for copper.

Cuprite *Cuprite* (copper oxide, Cu_2O) forms eight- or twelve-sided crystals of a striking colour which leads them to be called *ruby copper* or *red copper ore*. When mixed with iron oxide it has an earthy appearance, and is called *tile ore*.

Galena
(Lead sulphide, PbS.)

As might be expected, *galena*, also known as 'Lead Glance', can be easily recognized by its unusual weight and its dull colour, not unlike that of lead itself. It has a right-angled cleavage, is easily

Galena

broken into tiny cubes, and is very soft. It is the principal ore of lead and is found in Cornwall, Flint, Dyfed, Cumbria, Derbyshire and the Isle of Man.

Zinc Blende
(Zinc sulphide, ZnS.)

Zinc Blende, also called *blende* or *sphalerite,* owes its names—from the German and Greek respectively for 'deceiving'—to its often being confused with galena. Both crystallize in somewhat similar forms, though they differ in other respects. Blende is usually brown or black, and is then called 'Black Jack', but it may be yellow or almost white; its streak may be white or brown; its cleavage is conchoidal. It is fairly soft and it is, of course, much lighter than galena. The principal ore of zinc, it occurs in Cornwall, Dyfed, Derbyshire and Cumbria.

Tinstone
(Tin oxide, SnO_2.)

Tinstone is sometimes called *cassiterite,* after the name which the Phoenician traders of old gave to Britain, the 'Tin Islands' (Cassiterides). It is almost as heavy as lead but is harder. It is usually dark brown, but may be white or grey; its crystals have a very brilliant lustre and in its non-crystalline form its structure may resemble wood fibres.

It is the principal ore of tin and forms veins in and around the granite. The detached and water-worn lumps found in the beds of the streams and in the adjacent gravels are known as *stream tin*.

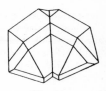
Tinstone

Pitchblende
(Uranium oxide ($2UO_3.UO_2$).)

Pitchblende—found in association with the tinstone of Cornwall—is so named because of its blackness and its general resemblance to pitch. It is exceptionally heavy and fairly hard. Its action in fogging photographic films led to the discovery of radium, as an impurity in the Cornish pitchblende. The mineral itself is also radioactive, being the principal ore of *uranium*. Formerly chiefly of technical interest, it has now become a substance of outstanding importance, on the proper use of which the future of mankind may depend, for one variety of uranium, isolated by a complicated and secret process, is used in the atomic bomb.

Wolframite
(An oxide of tungsten and iron or manganese, (Mn,Fe)WO_4)

The proportions of the two metals vary, from Hubnerite, manganese tungstate ($MnWO_4$) to Ferberite, iron tungstate ($FeWO_4$). Its crystals are tubular and have a perfect cleavage; its colour and streak are chocolate brown, and it may have a brilliant lustre or may be dull. It has been mined in Cornwall, where it is found along with quartz and tinstone. Steels containing tungsten, being very hard and strong, are used for high-speed cutting tools.

Alloys

A number of metals much used in everyday life or industry have no ores, for they do not occur naturally. These are the *alloys* (named from an Old French word meaning 'combine') formed by melting two or more metals together; the metals combine so completely that one might almost be dissolved in another. The oldest alloy may be *bronze*, formed of copper and tin. It is harder than copper, and does not corrode, its invention making improved tools and weapons possible; the modern *phosphor bronze* is exceptionally strong. *Brass* is an alloy of copper and zinc; *gun-metal* of copper, tin, and zinc; *pewter* of tin and lead; *bell-metal*, like bronze, is an alloy of copper and tin, but in different proportions. Alloys of mercury are *amalgams* (called from the Greek for 'softening') and have many uses; an amalgam of tin is used for 'silvering' mirrors.

Iron, as used industrially, is not pure but contains small quantities of carbon, which convert it into *wrought iron, cast iron,* or *steel.* There are many grades of steel, with their own special uses; many of them are alloys of iron with such metals as manganese, chromium, or tungsten. *Nickel steel* is especially strong and elastic.

SCENERY

The Mountains

Weather Action Upheaved by earth movements, or forced up in a molten state by pressure from below, the rocks, bedded or massive, rose high above sea-level. Even as they rose the weather began to destroy them.

By day, the sunshine beats on their surface, heats it and makes it expand. By night, the chill air cools it and makes it contract. Alternately stretched and compressed while the mass of the rocks remain unaffected, the surface is loosened, weakened, cracked and riven into boulders, fragments of stone, and grains of dust.

The rain beats on the weakened surface, washing away the smaller particles, dissolving such minerals as calcite, converting the hard feldspar into soft clay, leaving the insoluble materials as a residual dust, and seeping into the cracks.

The wind sweeps the weakened surface, driving the rain forcibly against it or blowing away the dust and hurling it against the rocks like a natural sand-blast.

The cold of winter freezes the rain in the crannies and makes it expand, slightly indeed but with irresistible force. Thus it continues the destructive action not merely on the surface but actually beneath it.

Scree Hardly is a surface layer removed when the weather attacks the new surface exposed. Thus it gradually whittles the rocks down. It strews their tops with loose boulders, and piles up on, and at the foot of, their slopes the loose material known as a scree. (It attacks the rocks elsewhere, with equally destructive effects, but in the lowland its action is delayed and concealed by the growth of vegetation and by the works of man.)

Convex Mountains It attacks the edges and corners of the rocks, exposed as they are to its action on two or three sides, more quickly than the flat surfaces. In this manner it smooths and rounds them, giving many mountains, as seen from a distance, a graceful convex outline.

River Action Flowing down the mountain sides, the streams carve them into valleys. Slowly widening and deepening these, at the same time they 'eat back' their heads, so as to rise a little nearer to the summits. The effect is to give a valley a head resembling the half of a hollow cone; this seems to have no English name, but in Scotland such a valley-head is called a *corrie*.

The streams on the two sides of a mountain ridge do not usually rise opposite one another. An 'alternate' arrangement, in which those on one side are roughly opposite the points midway between those on the other, is very common. As these springs eat back their heads they cut the crest of the ridge into an irregular zig-zag. Many mountain ranges and outlying spurs and ridges take such a form.

The springs on a compact mountain mass rise at irregular intervals round its slopes. The valleys they form at last become so deep and wide, and trench so deeply into the central mass, that they cut this into a series of ridges extending outwards like the spokes of a wheel. This result is well seen in the Lake District, Cumbria.

Concave Mountains Even as the river valley deepens, its sides are destroyed by the weather. The combined action, unless the rocks are exceptionally hard, gives them a concave shape as graceful as the convexity which the weather gives to the mountain slopes. It even extends to the tops of the ridges themselves, so that a series of such curves, sweeping from one summit to the next, forms the typical skyline of a mountain range.

A mountain, deeply trenched by the rivers on one side but weathered on the other, may combine both curves. Its outline then somewhat resembles that of a sand-dune—or an ocean-wave.

These results are, however, modified by differences in the rocks. A soft rock weathers in curves; a hard rock is more apt to split along its bedding or down its joints, producing plateaux more or less level bounded by vertical cliffs.

The effect of these varied influences on no less varied rocks is to overcome any tendency to a monotonous regularity. They give every range and peak an individuality which to the poetic mind almost suggests personality.

River Valleys

Formation of Valleys   The rivers trench the hills into valleys not so much with the water itself as with the loose material it sweeps along. In the fine summer weather, when they are reduced to a gentle trickle, they seem quite ineffective. But when the downpours of winter and the melted snows send them roaring down the hillside, the destructive work is very plain. A heavy snow, a sudden thaw, or a cloud-burst, may so swell them that they visibly destroy great masses of earth. Their customary action, winter after winter, though less spectacular, is far more destructive.

Pot-holes   It is obvious that a rapid stream, flowing through soft rocks and scraping large pebbles or jagged lumps of stone against them, will deepen and widen its bed. It seems less probable that it can similarly destroy even the hardest rocks. One interesting stage in the process is the formation of *pot-holes*.

These are rounded hollows in the river-bed, their sides as smooth as the surface of the pebbles which most of them contain. They were formed, and are intermittently enlarged, by these pebbles, which, spun round by eddies in the water, act like boring tools and grind holes in the rocks. They may be enlarged by this action until they are separated only by narrow walls, easily smashed by the other stones which the torrent carries downstream. This lowers the whole level of the river-bed.

Chemical Action The water itself not merely washes away the loose materials over which it flows. It dissolves them and also destroys them by chemical action. For it is by no means 'pure' water; it is a solution of carbon dioxide from the air, calcite from the limestone and chalk, organic matter from the peat-bogs, and casual impurities. Some Scottish streams attack the limestone so powerfully with material derived from the peat-bogs that they wear away the sides of their channel, leaving their tops overhanging.

River Capture As it widens and deepens its channel and 'eats back' its head, a river may trespass on the valley of a neighbour. Whichever of the rivers has the lower channel will 'capture' the water of the other and divert this into its own channels. It may reverse the flow of a whole stretch of water; it may drain off a channel and leave its valley dry; it may reduce a large stream to a comparative trickle. The disproportion between the present volume of part of the Upper Thames' tributaries and the valleys through which they flow may be due to the interception of their headwaters by the Severn. A clearer example of river capture is given by the Medway; originally one of a number of short streams flowing northwards through gaps in the North Downs into the Lower Thames, the head of its western tributary 'ate' its way back through the soft Gault Clay until it intercepted and captured the headwaters of the Darent.

River Scenery A river's speed depends partly on its volume and slope, partly on the nature of its bed. It destroys soft rocks quite easily, widening and deepening its course at the expense of its speed. Among hard rocks, it can cut only a shallow narrow channel, through which it rushes violently. Alternations of soft and hard rocks produce a correspondingly varied scenery.

A gradual uplift of the land will increase the speed of a river, making it attack the rocks with renewed vigour. A stream which hitherto has been meandering sluggishly to the sea may thus cut deeply into the rocks below. It will then produce an unusual type of scenery, a gorge as narrow and craggy as those which cut direct through the mountains but as sinuous as those which wind their way across the plain. By this process the Wye cut its deep windings near Symon's Yat.

Waterfalls

A river, in flowing from a hard rock on to a soft one, cuts a deeper groove in the soft material than in the hard. Thereby it produces a 'step' in its bed, and in falling over this step it acquires greater force and cuts its channel even deeper. The step becomes a *cascade* over which the water splashes if its face be jagged; a *waterslide* if it slopes gently; and a *waterfall* if it be upright. Below the step the river correspondingly increases its speed, forming the *rapids* which can be so dangerous near a really large fall.

The higher the fall, the greater the impact of the water and the deeper the pit it cuts. Its swirl in the pit meantime cuts away the banks, while the loose material it brings down piles up just below the pit to form a natural dam. At the foot of many waterfalls, therefore, is a deep pool, almost circular in shape and overshadowed by steep banks.

The falling water gradually wears away even the hard rock-face down which it drops, and may at last produce a shallow cavern. In these islands such caverns are quite small, the waterfalls which produce them not being very powerful or high. A really large fall, however, can produce so spectacular a cavern as the 'Cave of the Winds' behind Niagara.

Lakes

Where a river crosses from soft beds to hard, its channel is suddenly constricted both in width and depth. The water, its flow thus checked, may spread and extend over the low-lying country behind the barrier to form a lake.

This is by no means the only method whereby lakes can be formed. A stream may be dammed by the scree fallen from a mountain-side, by the loose material banked up long ago by a vanished glacier, by the sediment brought down by a larger river, by a shingle-bed cast ashore by the tide, by a barrier artificially built to make a reservoir. Earth movements or the scooping action of a glacier may form a depression in the ground, or extensive mining operations may cause it to subside, and the streams or rainfall may fill it with water.

Ripples on a lake's surface may act, though of course on a smaller scale, like the waves of the sea. They may produce a miniature cliff, or a sandy beach with its characteristic regular curve. Elevations in the bed of the lake, or masses of loose material, may produce islands. Tangles of loose vegetation, covered with silt, and buoyed up by the gases produced as the vegetation decays, may form 'floating islands', and as the gases escape these may sink.

The Lowland Plain

Deposition of Sediment The general course of a long river is concave, steep near its source but levelling out towards its mouth. Here its slope may be barely steep enough to keep its waters in gentle motion, and under such conditions its effect on the landscape is the reverse of that produced by its headwaters. It now forms more land than it destroys.

The stream is already laden with all the sediment it can carry, and will deposit some of this whenever its flow is checked. When it overflows its banks, it carries the sediment with it; but when the floods subside they leave a deposit of silt spread over the land. The silt fills the depressions in the ground, and has a general levelling effect, but it may also accumulate in low ridges along the water's edge.

Changes of Course We are apt to think of rivers as having courses as clear-cut and permanent as the lines which represent them on the map. Unless, however, a river is properly embanked and carefully watched, it will intermittently change its course. It may do so suddenly when a sudden spate breaches a gap in its banks and finds a new course towards the sea. Or it may do so more gradually, by a process almost imperceptible to the casual observer.

A very small obstruction will divert its flow, directing it away from one bank and towards the other. Once started, the process is cumulative, for the stream washes away the bank towards which it is diverted and deposits sediment at the foot of the other.

Where the stream curves, its outer bank rises steeply and may even be undercut—until its lip collapses, to be swept away by the water. Its inner bank slopes more gently and may end in a long 'split', running out into the stream. At an 'S-shaped' curve part of each bank will rise steeply; and two splits may run out, one from each bank, to produce a ford.

Abandoned Meanders The process does not end here; the stream goes on attacking the outer bank and depositing silt on the curve's inner side. Thus the curve gradually becomes more pronounced, until it may constitute the greater part of a circle, 'doubling back' on itself with only a narrow strip of land separating its beginning and end. Finally this strip may be destroyed, so that the water no longer flows round the curve but cuts right across it. The old channel, choked at both ends by sediment, forms a narrow horseshoe-shaped 'ox-bow lake', until its water dries up, leaving nothing but a horseshoe-shaped dell. Such 'abandoned meanders' may be seen in many valleys, showing where the rivers once flowed.

Deposition of Silt These changes of course cover the floor of a river-valley with a level deposit of silt, and beside a large river produce a wide expanse of plain stretching inland from the sea. The best-known example of such a plain is, of course, continental Holland, produced by the Rhine. But England has its own Holland, the southern part of Lincolnshire bordering the Wash, produced by the Ouse and the other rivers of the Midland Plain; and the triangular strip of flat country pointing westwards from London was formed by the Thames.

A river whose flow has somewhat increased through an uplift of the land will destroy the very plain which it formed. Normally, however, the destruction will not proceed far enough to reach the bed-rock below the accumulated silt. Presently it is checked, and the river forms a new plain at a lower level; further earth movements may cause the whole process to be repeated, and the river may form a third plain still further below.

River Terraces The destruction of the older plain is, however, not complete. It may leave low, flat-topped hills in the middle of the plain, and narrow strips along the hills at its edge. These strips are called *river terraces,* and may be traced on the hill-sides bordering most large river-valleys. The Thames has three clearly marked terraces: the lower, only about seven metres or so above the present high-water mark, borders the river at, for example, Battersea on one side and Chelsea on the other; the second, about fifteen metres above the water, includes Hyde Park and Putney; the third, at about thirty-three metres, includes Tooting Common and Pentonville. These terraces mostly consist of gravel and have yielded many interesting relics of early man.

Formation of river terraces

The Cliffs

Destruction by the Sea The coast is exposed to the destructive force not only of the weather but of the sea. The waves pound it with countless tons not merely of water but of rock-fragments and pebbles, and rasp at its surface with thousands of tons of gritty sand. They rush into its crevices, compressing the air within them, and then retreat, allowing it to expand violently. Thus, like the weather, they destroy the rocks not only on the surface but a little below; and, like the weather, they have hardly destroyed one surface before they begin on another.

The waves destroy the rocks at, and somewhat above, sea-level, undermining the cliffs and making their summits collapse by their own weight. The weather acts more generally, and even more destructively, wearing away the whole face of the cliffs. A layer of loose fragments may, though only for a time, protect the crests and slopes of the hills against destruction by the weather; it does little to protect a vertical cliff.

The speed of the destruction of course depends on local conditions; the strength of the tides and currents, and especially of the wind by which the waves are produced, is as important as the character and build of the rocks. Even the great Atlantic breakers can destroy the Land's End granite but slowly; the lesser waves of the Channel are attacking the softer chalk cliffs so effectively that to the east of Brighton these have had to be elaborately armoured with concrete; the North Sea is visibly sweeping away the East Anglian shell-beds and sands.

Formation of Cliffs   Where one type of rock extends for some distance, it forms a long stretch of cliffs, diversified by local differences either in itself or in the sea. Joints and veins of softer minerals produce creeks and coves; masses of unusually hard material become islands and reefs. At or above sea-level, the waves may bore openings large enough to be regarded as caves, though small in comparison with those in the limestone. A tapering cave may pierce a wall of rock, forming a blow-hole through which the waves spurt with an alarming or amusing sound.

Bays and Headlands Where soft and hard rocks lie side by side, the sea destroys them at different rates, producing a varied coastline of alternate bays and headlands. Where a soft bed lies a little inland behind a cliff formed of hard rocks, the destruction is slow until a hole is breached in these, then becomes much more rapid. Stair Hole, near Lulworth Cove, Dorset, is a result of such a breach in the limestone cliffs, the clay behind having been destroyed to leave a funnel-shaped pit at the bottom of which the sea roars and foams. At Lulworth Cove itself the cliff has been cut completely through, and the clay has been 'slipped' seawards, leaving a circular bay. Farther west, the limestone has been reduced to a line of reefs (the 'Man o' War Rock', the 'Cow and Calf' and others, including one joined to the mainland by a rock-arch, the 'Durdle Door'), the clay has been completely destroyed, and the sea is attacking the chalk. Farther east, miles of the chalk have been completely swept away, separating the Isle of Wight from the mainland.

Land-slips Where a varied series of rock-beds, especially if they include clays, lie more or less level, the action of the waves may be catastrophic. Their base gradually worn away, the cliffs suddenly collapse into the masses of confused material known as *land-slips*. Consisting as they do of a variety of rocks, these produce a rich soil which fosters a tangled vegetation, the haunt of numerous adders. The Warren between Folkestone and Dover, the Land-slip west of Lyme Regis, and the Undercliff west of Ventnor in the Isle of Wight, were all produced by material fallen from the cliffs whose faces now rise some distance inland.

Drowned Valleys and Raised Beaches   Not merely is the land destroyed by the sea. Evidences of its having sunk bodily are given by the fossil-forests already alluded to and by such *drowned valleys* as those of the Dart, Fal, and Exe, unmistakably different from the ordinary inlets and with unmistakable resemblances to the mountain valleys artificially 'drowned' to form reservoirs. Elsewhere there are evidences that the land has bodily risen: *raised cliffs*, with unmistakable beaches at their foot, some distance above high water mark.

Between the Tides   The appearance of a rocky shore is familiar. Above high tide the rocks are jagged; below, they have been smoothed and their angles rounded by friction with the wave-born pebbles and sand. Their depressions form delightful rock-pools, the home of interesting forms of marine life. Some of the smaller pools are almost circular and contain rounded pebbles; they resemble the pot-holes of a river and were similarly produced, by friction with the revolving pebbles whirled round by eddies in the water.

The Beaches

Sea-borne Materials Like the rivers, the sea not only destroys the land, it uses the fragments to build new land elsewhere. Grinding them together, it smooths them into rounded pebbles or reduces them to grains of sand. It carries them along the coast and throws them up on shore. It washes them off and throws them on shore repeatedly, until at last they attain a temporary rest, perhaps miles from the cliffs they once helped to form.

The faster the sea moves, the larger the materials it can carry; and it moves faster when the tide co-operates with the currents than when it opposes them. In spite of the ebb and flow of the tide, therefore, there is a general movement of the pebbles and sand along the coast, up-Channel, for example, and down the North Sea.

Groynes This movement has to be checked by the *groynes,* vertical walls of timber or masonry running sea-wards down the beaches. Its existence is well shown by the difference in height of the shingle on the two sides of the groynes.

Shingle Beaches A cape, or island, may act like an immense groyne, checking the movement of countless pebbles and making them pile up in a great ridge. Portland Bill has thus produced Chesil Bank, a shingle-beach forming a graceful curve several miles long joining the island to the Dorsetshire coast. Other banks of shingle are formed where the pebbles are swept directly on shore.

Permeable though the shingle is, the water seeps through it but slowly. It therefore dams the larger streams, producing marshes, ponds or

The River Alde

lakes, some metres from the sea. The Chesil Beach has produced a whole line of such ponds. In Cornwall the shingle has dammed the Hele stream, below Helston, to form a very beautiful lake. In East Anglia the partial damming of the

River Waveney brought into being the Oulton Broad.

The river may find its way to the sea round a shingle bed. Thus the River Alde once reached the sea near Aldeburgh but was diverted southwards by a growing stretch of shingle formed by the currents of the North Sea. It reached and passed Orford, eight kilometres down the coast, and now flows parallel to the shore for sixteen kilometres before it reaches the sea.

The currents may pile the shingle up not on, but perpendicular to, the shore, forming a *spit* running some distance out to sea and ending in a shape like a hook. The spit at the western end of the Solent extends from the Hampshire coast half-way to the Isle of Wight, and is so firm that Hurst Castle is built on it. The spit which extends from Spurn Head is slowly lengthening southeastwards into the North Sea.

A barrier to rivers, the pebble-beaches protect the coast against the ravages of the sea. So valuable are they that on many beaches it is forbidden to cart away the shingle; its large-scale removal would allow the waves to destroy the land.

Coastal Curves As the speed of the water slackens, it drops first the larger pebbles, then the smaller, and lastly the sand and silt. Thus it 'sorts out' the material it carries; on many beaches the pebbles lie apart from the sand, and are roughly graded according to their size. The action of the sea also piles up the material it throws ashore in graceful formations, of which the largest is perhaps Cardigan Bay, running from cape to cape.

Where there are no hard rocks to form capes, on the other hand, the curve, though no less graceful, is convex, as may be seen on the East Anglian and Lincolnshire coasts. Broken though they are by river-mouths, the general outlines of their curves are unmistakable.

Quicksands   Where a current is checked, either by an opposing current or by the sea, it drops much of the material that it carries. Dangerous shoals and sandbanks thus form in the sea, and harbours and river-mouths have to be elaborately dredged to keep them from being blocked. Especially dangerous are the *quicksands,* stretches of waterlogged silt which, though apparently firm, suck down and smother anyone rash enough to venture on them.

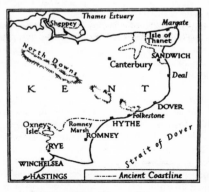

Map of the Cinque Ports

New Land Material dropped by the sea may clog the channels between an island and the coast, and convert it into part of the mainland. Thus the 'Island' of Thanet, now part of Kent, was once separated by a narrow channel so deep that sea-going ships took advantage of its sheltered waters. Still more surprisingly, an inland hill was once separated from Kent by a still narrower channel; it still keeps its old name, the 'Island' of Oxney.

The sea-dropped material may also clog inlets, and form wide stretches of coastal plain in the lee of the headlands. Pevensey Level, and other recently-formed coastal plains on the shore of Kent, have cut off from the sea several of the Cinque Ports once responsible for the marine defence of England.

Moorlands

Rocks that form a soil unsuitable for cultivation produce stretches of moorland, 'forests' in the historical sense. On sands and sandstones they are very dry and consist of pinewood and heath; on granites they are dry in summer but marshy in winter, and are clothed in heather and furze. On certain other rocks they are permanently marshy; many plateaux (elevated plains) and the flat tops of mountain-ridges form such scenery. Here the springs seep or bubble out of the ground, but as there is no distinct slope down which their water can flow, it forms a maze of channels broken by irregular banks of earth and patches of marsh.

Much of the soil consists of peat, and the water is stained brown by the decaying vegetation and yellow by the bog iron ore (limonite). Only when it reaches sloping ground does it become clear and form a definite channel. Robert Montgomery, a nineteenth-century rhymester, was trounced by Macaulay for saying that 'streams meander level with their fount'; on level stretches of moor this is exactly what they do.

Glacial Action

Glacial Drift   The scenery south of the Thames differs subtly from that farther north; for example, the grass-clad Downs contrast strikingly with the wood-clothed Chilterns, though both are formed of chalk. So great is the difference that whereas in the south one series of geological maps is sufficient, elsewhere two are necessary: the *drift* edition showing the surface material and the *solid* edition ignoring this and indicating the underlying rocks.

This difference is due to the Great Ice Age, that period of intense cold, hundreds of thousands of years long, which ended about fifty thousand years ago. The polar ice then extended into Central Europe and covered England as far as the Thames, and its movements affected the scenery as do the Alpine glaciers.

Rock-flour   The moving ice ripped away the surface not only of the ground but of the actual rocks. It crushed many of the pieces into a powder so fine that it is called *rock-flour*, mixed this with earth and water, churned the whole mass together, buried rock-fragments in it, and so produced the Boulder Clay. It spread this over the land, partially filling the depressions and smoothing and levelling much of the scenery.

Roches Moutonnées The ice rubbed the edges off the larger boulders, rounding and smoothing these and giving them a resemblance to the backs of crouching sheep. Their name, *roches moutonnées*, comes, however, not from this but from an equally fanciful resemblance to once-fashionable wigs! The smoothing may take place on three sides only, that away from the approaching ice remaining rough, and so indicating the direction in which the glaciers moved.

Glacial Striae The ice similarly smoothed the faces of the cliffs. They, like some of the *roches moutonnées* and of the detached fragments in the Boulder Clay, bear parallel grooves and scratches formed by the gritty material which the ice rasped against them. These *glacial striae* indicate not only the direction in which the ice moved but the height it reached, marked by the striking contrast between the smooth striated rocks below and the jagged cliffs above.

Hanging Valleys The glaciers, in advancing down the valleys, ripped off the tips of the spurs extending from the hills on either side. They thus straightened the general course of the valleys, enabling a clearer view to be obtained along them. They deepened them, so that tributaries join the main river not on a level but by cascades flowing down the sides, producing *hanging valleys*. They also altered the slope of their sides; an ordinary river valley is *V-shaped*, a glaciated valley is *U-shaped*.

Moraines The glaciers carried much loose material with them, piling it into crescent-shaped ridges called *terminal moraines;* in the Pennines, in Central Wales, and in the Lake District a number of valleys are dammed by such moraines, behind which the rivers expand into lakes. The glaciers also carried large boulders a great distance from the rock-beds from which they were ripped. These are called *erratics* because they are of a different type from the rocks among which they now lie; and when, as in Llanberis Pass, they stand in places apparently inaccessible, they are called *perched blocks*.

Rock Basins and Corries A glacier has a power unlike that of any river, of scooping depressions in the ground, and even in the rocks, over which it moves. The floor of a glaciated valley may thus rise or descend abruptly in 'steps', and the depressions formed, called *rock basins,* are likely to fill with water and become lakes. Loch Coruisk, in the Isle of Skye, so impressively overshadowed by the mountains that some visitors find it unnerving, is a rock basin whose floor is somewhat below sea-level; Glaslyn and Wastwater are also rock basins, one high on the shoulder of Snowdon, the other in the Lake District. The steep-sided cones scooped into the mountains by the ice at the head of the glaciers are known in Scotland as *corries*; they seem to have no English name.

Human Handiwork

What is 'Nature'? The effects of human handiwork are smallest in the mountains: mines, quarries, communications (railways and roads) and a few dams. They are localized on the shore in ports, seaside towns, fishing-villages and odd houses and bungalows in the sheltered coves. They are sporadic in the forests, whether the word has its traditional or present-day use, in the moors and woods. They are practically continuous over the level or undulating country where the rocks produce a fertile, easily cultivated soil.

The townsman is apt to think of 'the country' as 'nature'. But, as H. G. Wells pointed out years ago, a ploughed field is as artificial a thing as there is: it can be produced only by means of appliances artificially constructed (nowadays by a complicated industrial process) through age-old human skill used for human ends; it is like nothing in nature; and left to itself—as every suburban gardener will realize—it quickly loses its character and becomes a tangle of grassland and weed.

All cultivated land is artificial. Pasture land more resembles untouched 'nature', but even this is kept unusually cropped by artificially introduced and protected cattle and sheep, and some of it demands artificial ditching. Much of the rich agricultural country of East England, to all appearances 'natural', was formed artificially from a water-logged swamp by the draining of the Fens. Wicken Fen remains to show this region in its 'natural' state.

Ornamental park-lands are equally 'artificial' —a word that formerly had no derogatory sense but simply meant 'in accordance with the rules of art'. Much of the country was deliberately planned by such landscape-gardeners as Capability Brown. Coppices were planted; streams were deflected into their present channels or dammed to produce lakes; impressive vistas were deliberately arranged. The waterfall in Windsor Great Park is as 'artificial' as the synthetic ruin besides which it flows.

Man as a Geological Agent Such alterations might be considered unimportant geologically because they affect only the surface of the ground; the same might almost be said of the glacial drift. Other works of man are, however, undeniably of geological significance. A reservoir in the mountains resembles any other lake except that its waters are restrained not by a ridge of rock but by a concrete dam. A whole range of hills, the Malverns, formed of a hard igneous rock very suitable for road-making, is in danger of being bodily quarried away. The amount of slate excavated from the Welsh mountains would have taken ages of weathering, or catastrophic earth movements, to destroy it. A railway and a line of shipping may be as effective as a large river in transporting material from the hills to the sea. A canal acts like a slowly moving river with an unusually regular course. The concrete facing of the Brighton cliffs has protected them from destruction by the sea as effectively as would a basalt dyke.

Important effects have also resulted unintentionally. Too intensive a cultivation of the corn-lands of America so loosened the soil as to convert them, in part, into a 'dust-bowl' from which clouds of grit blow, with devastating effect, over the adjoining farms. (Similar ill-considered action, in antiquity, may have produced some of the earth's deserts.) Wholesale destruction of the North American forests to yield pulp for cheap newspapers is threatening not only to destroy the soil but also to affect the climate adversely, for a tract of woodland gives off large volumes of water-vapour.

Re-afforestation, on the other hand, though undertaken for the sake of timber, may 'bind' the soil and ameliorate the climate. Large-scale irrigation may reclaim deserts by converting them into fertile land. The most grandiose scheme so far suggested—one would hesitate to say that it is impracticable—is to melt the polar ice-caps by means of atomic bombs, an operation which would certainly have far-reaching and incalculable geological results.

208

211

GEOLOGICAL MAP OF THE BRITISH ISLES

Recent
Pliocene
Eocene and Oligocene
Cretaceous
Jurassic
Triassic
Permian
Carboniferous
Devonian
Silurian
Ordovician
Cambrian
Torridonian
Pre-Cambrian, etc.
Granites, etc.
Lavas, etc.